Polaris

Lessons in Risk Management

Dr. John J. Byrne, PMP

First Edition

Multi-Media Publications Inc.
Oshawa, Ontario

Polaris: Lessons in Risk Management
by Dr. John J. Byrne

Managing Editor:	Kevin Aguanno
Typesetting:	Charles Sin
Cover Design:	Charles Sin
eBook Conversion:	Agustina Baid

Advisor: Franklin Knemeyer, Deputy Technical Director, US Naval Weapons Station China Lake, CA (Retired)

Peer Reviewers: Thomas Smith, PMP, LTC US Army (Retired), Technical Project Manager in the Information Solutions Department of the Keystone Mercy Health Plan

Russell J. Figueira, PMP, RMP, MSM, MPM, Vice President of Professional Education American College

Adam Selverian, PMP, RMP, MSA, Director Release Management, IKON Office Solutions

Published by:
Multi-Media Publications Inc.
Box 58043, Rosslynn RPO
Oshawa, ON, Canada, L1J 8L6

http://www.mmpubs.com/

Paperback	ISBN-13: 978-1-55489-097-2
Adobe PDF ebook	ISBN-13: 978-1-55489-098-9

Published in Canada. Printed simultaneously in the United States of America and the United Kingdom.

CIP data available from the publisher.

Table of Contents

Dedication

For Jerry, Joe, Tom, Mary, all those that served on Polaris
Submarines, and my loving wife, Wanda.

Polaris: Lessons in Risk Management

Acknowledgements

When I started this project oh so many months and years ago, I never envisioned getting to this point and actually finishing the book. The elephant seemed too large to ever finish eating. There were so many hurdles to be overcome. Without the assistance from the below individuals, I doubt I would have completed the book.

In addition to providing me with much needed motivation to finish the book, Jim Snyder, wrote a wonderful forward for this book. Jim Snyder is one of the founders of the Project Management Institute and is a Project Management Institute Fellow. I have worked with Jim for a number of years now on various educational projects. Each project has been a pleasure for me to have the opportunity to work with Jim. Jim is a great visionary. I look forward to many more projects with Jim. There is little chance this text would have even gotten published without his helpful motivation.

Next, I would like to thank Russ Figueira. Russ provided guidance and assistance through many facets of the book's development. He was always just an email away should I need a second or third or fourth or fifth reading of the book. He assisted most in ensuring the risk management portion of the book would make sense to the reader and was factually correct. He constantly went out of his way to help me in any way possible with regards to this book. He made time in his busy schedule whenever I needed it.

Tom Smith was the man behind the scenes asking "Is this what you really wanted to say- why did you write it that way?" Or, he would comment, "You better add this into the book. It just does not make sense without it." He could always find those little missing pieces to make the book better and more readable. Being a retired U.S. Army Lieutenant Colonel, may have allowed Tom to do this easily. I don't know, but I do know he is good at it.

Maria Fama is not a project manager nor is she an expert on the Polaris Project. Instead, she is an English professor. She reviewed the book from a different viewpoint than all the other reviewers. She read the book from the viewpoint of an individual with no preconceived notions about the book or its content. This was invaluable in finding gaps and other issues in the text. Without her assistance, I am not sure the book would have been readable.

Adam Selverian provided a technical peer review of the risk management portions of the book. He was able to find a number of technical mistakes I had made, thus improving the quality of the book.

Frank Knemeyer is a retired engineer from China Lake Naval Weapons Station. I was very, very lucky to find him. Frank was an engineer on the Polaris Project. Some of the work described in this book and accomplished by the Polaris Project was his handiwork. His first hand insight into the Polaris Project added more to the book than any other single entity excluding myself. The book went from an "all right" book to the book it is today by Frank's assistance. If all of the

Polaris Project's engineers were like Frank, I can understand why this project was successful.

China Lake Naval Weapons Station Museum and China Lake Alumni Organization – Gary Verver, were very helpful in the construction of this book. Both were quick to provide all the photos associated with the Polaris Project they possessed for use in the book. These photos added much to the realism of the book. These pictures brought the project to life. China Lake Museum provided one more extremely important piece to this book: Frank Knemeyer. Without the assistance of the museum, I would have never found Frank and the book would have suffered accordingly.

Continuing, I would like to thank the Lessons from History Series, especially – Mark Kozak-Holland, and Multi-Media Publications for the opportunity to have this book published.

Lastly, and most importantly, I would like to thank all of my students that have been pushing me for years to write this book. Class after class has clamored for me to get behind the computer and start typing. I finally listened.

My personal interest in this project started when I was stationed on the U.S.S. *Nathanael Greene* (SSBN 636) as a nuclear reactor operator. The *Nathanael Greene* was a Polaris submarine. Having spent many a month on board the *Greene*, it was easy to wonder while walking around the ship how all of this was possible. How were they able to create such a machine, such a weapon? After doing the research for this book, I am amazed at what they accomplished. Today the sail of the *Nathanael Greene* sits on display at Port Canaveral, Florida.

– John J. Byrne

Foreword

Polaris: *Lessons in Risk Management* is an exciting little book about a very, very big subject. Dr. John J. Byrne takes on the task of defining the risk management process in terms that make it not only understandable but interesting. He uses a milestone in history; the development of the Polaris missile and its submarine based launch system project, as the setting in which to develop and explore the processes and application of risk management within modern project management techniques.

The birth of the Polaris Missile Project and the SSBN Submarine Project was directly related to the Russian launch of *Sputnik* in October, 1957. This event was a threat to US strategic defense and a response needed to be quick and decisive. The project gave birth to PERT and, along with James Kelley's development of CPM for DuPont, this project saw the start many of the concepts of modern project management. The management of risks associated with this project was critical to successful completion. Warren Buffet has been quoted as saying that "Risk comes from not knowing what you're doing" and it is certain that we did not know what we were doing at the start of the Polaris project.

Dr. Byrne approaches this subject from the prospective of one with firsthand knowledge of the project and concepts involved. As a nuclear reactor operator who served on two nuclear submarines, he is uniquely qualified to discuss the project and its associated risk management. By using this fascinating response by the US Navy to an international threat, Dr. Byrne makes a difficult concept interesting and understandable.

After setting the historic stage for the Polaris project, Dr. Byrne defines a six step risk management process and then breaks each part of the process into its components describing them in detail and relating them to his framework project – Polaris. This straight forward approach to defining and understanding risk and risk management makes clear each of the process components and illustrates their importance to successful project management.

Dr. Byrne concludes his book by discussing the successful completion of the project and the people responsible. I am sure you will find this book extremely useful in understanding and applying the risk management process to your projects as well as a joy to read . Don't stop now, keep going at full speed ahead!

– James R Snyder, Founder and Fellow, PMI

Introduction

The Project Management Institute (PMI) reports that $15 trillion in U.S. dollars is being spent worldwide in 2010 on projects. That number represents 1/5 of the world's gross domestic product (GDP). That is one impressive number! Even with the world economy in the Great Recession that much money spent on projects is truly mind boggling. One would expect with that amount of money being spent that the success rate of projects would be high, if not even very high as would investment in project managers and project management.

With that much money being invested, undoubtedly projects account for huge monetary outlays for corporations. It would be reasonable to assume, with that amount of investment at stake, companies would be ensuring their project managers have the skills, knowledge, and abilities to bring projects in on time and on budget. It would also seem reasonable to assume that a portion of the money being spent on projects would be at least partially spent on training project managers.

Unfortunately, little of that is true. According to the Standish group, only 32% of projects were successful last year--- 32%! That's about one in three projects successful! Further, this is noted to be the lowest success rate in the last five years. I doubt many professions other than possibly a baseball player could be considered successful with a 32% success rate. This seems unconscionable considering the trillions of dollars at stake worldwide.

With project success rates that poor, just think of the money that is being wasted each year on poorly run projects! I have read reports stating that between $260 and $285 billion dollars was wasted due to project failures in the Unites States alone last year. That is over a quarter of a trillion dollars lost in the United States alone! Imagine what that value is if extrapolated worldwide.

Sadly, part of the poor success rate and excessive money lost due to failed projects can be traced to a lack of employing project management methodology and a general lack of providing training to project managers. This is unfortunate since numerous studies have shown that the use of project management methodologies improves project success rates, adherence to budget, adherence to schedule, and overall project efficiency. Moreover, the Project Management Institute reports that there are approximately 22 million individuals worldwide with the job title of project manager. Less than ½ a million of them have had any formal training in project management. Additionally, a 2004 Carbone and Gholston Study reported that less than one in four project managers have had any training, formal or otherwise, in project management. This study further reported that 85% of all project managers have had four days or less of training in project management in their entire lifetime. This is surreal. Imagine that your doctor had 4 days of training in medicine and a success rate of one in three. How long would you continue to see that doctor? A similar question can be asked of project management – why doesn't industry spend more on the training of project managers and why don't individual companies work to improve their success rate on projects? It would likely improve their bottom line. This is certainly an interesting point of discussion.

As one can see, it appears many project managers need to learn their trade on their own. One area in particular within project management that receives, in my opinion, the least emphasis in training and in use is risk management. I have asked why this is so to many in industry. I get many different answers. Some say it is just too difficult to do. Some say it is too complicated to learn. Some say books on the subject are difficult to read and comprehend – these books being filled with statistics and complicated equations. Some simply say it does not work. Some, incredibly, simply say they never heard of it. Some say they don't need it; they have enough to do putting out fires on their projects.

This book was written to show that, for one, risk management does work and, for another, it is neither difficult to learn nor to use. This book utilizes the Polaris Submarine Program as an example of a successful utilization of risk management processes. The Polaris Missile project was considered by many as impossible to complete without the use of project risk management techniques such as Program Evaluation and Review Technique (PERT) and other, now common, risk management techniques. These developed on the Polaris Project are precursors to modern day risk management techniques. Some of today's risk management methods were developed solely for this project.

Risk management is a process for managing uncertainty. Uncertainty is a key ingredient in every project, as many aspects of a project are uncertain until they either happen or they do not happen. Whether a critical component will arrive on time is uncertain until it actually arrives on the job site. Whether a key resource is available may not be known until the time he or she is to start work. This is true regardless if we want it to be or not. Failure to properly manage risks associated with a project almost certainly dooms the project to poor performance. Risk management provides a structured methodology to manage this uncertainty/risk by identifying risk, analyzing risk, proactively treating significant risk, and monitoring and controlling risk throughout the life of the project.

The Polaris, Fleet Ballistic Missile (FBM), or SSBN program, as is was also known, will be followed throughout the book as a method to visualize risk management in action. In this way, one can learn risk management, or improve one's existing knowledge, while seeing how risk management techniques are actually used, not just described as in many other books.

This book does not cover developing requirements, developing a scope statement, or creating a Work Breakdown Structure (WBS). Nor does it include activity generation, budget generation, quality management, communication management, or scheduling. Instead it considers these other important processes of a project to be in progress or completed, as this text concentrates on risk management solely.

This text describes modern day risk management techniques. The Polaris Project in many cases did not employ what we would call "modern" risk management techniques to solve the risk related issues of the Polaris project, as many of these techniques and terms did not exist at the time. Hence, this text describes how such risks would be attended to in today's risk management environment.

This text is not designed to help you pass the Project Management Professional (PMP) examination, though it may be very helpful in that regard. Nor is it designed to help you pass PMI's Risk Management Professional Certification, though it could probably help there also. This book is designed to provide a straightforward approach to risk management that can be employed by any practitioner of project management. This book should serve as a ready reference for all project managers and project risk managers. This book is based on lessons from history.

Polaris: Lessons in Risk Management

## CHAPTER ONE

Sputnik

O ctober 4, 1957 was the date that changed project risk management forever. That was the date project risk management got kicked out of its doldrums by a little round shiny object. On that date the Soviet Union launched Sputnik. It was the world's first artificial satellite. Sputnik was a 184 pound satellite, insignificant by today's standards. It was a little less than 23 inches in diameter, roughly the size of a beach ball with the weight of a well built man. Not really much of an imposing figure compared to modern satellites. The only thing Sputnik was designed to do was to simply emit a beeping signal – that's all. But this 184 pound, 23 inch manmade object launched a space race and arms race between the USSR and the United States, introducing the world to the Space Age – an arms race and space race that would last for decades and cost trillions of dollars and trillions of rubles.

Sputnik was a stunning scientific and engineering achievement for its time. At that time the United States was designing the Vanguard Rocket as its largest rocket. The Vanguard Rocket was to have a designed payload of 3.5 pounds. This is nothing compared to the payload of Sputnik. Sputnik was a bigger rocket, could carry a larger payload, and fly higher than the Vanguard. Simply put, it outclassed the United States' best.

America was caught completely off guard by this outstanding scientific and engineering achievement. Prior to this, the United States considered itself the only country capable of such a scientific and engineering achievement. Hence, this satellite launch created an uproar at all levels in the United States – from the person in the street to the president in Washington. A sort of "Sputnik Mania" overtook the people of the United States. Some individuals characterized this "Sputnik Mania" as bordering on hysteria. The psychological effects on the United States were profound! Thus began the days of "duck and cover" drills in the nation's schools.

Why did this single event cause so much uproar and hysteria in the United States? Fear – that's why! To begin with, it was quickly realized by the government and the general

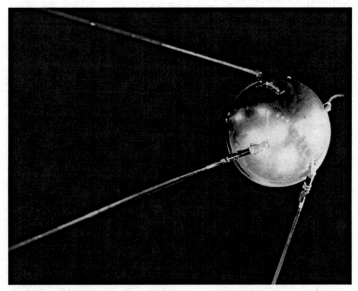

Figure 1.1: Sputnik I, courtesy NASA

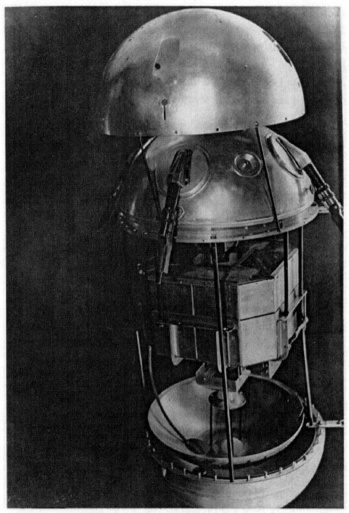

Figure 1.2: Sputnik I, courtesy NASA

public in the United States that if the USSR could launch
Sputnik into space, it could also launch missiles with nuclear
warheads aimed at the United States instead of a simple
beeping scientific satellite. This could quickly decapitate the
government and render the United States' bomber fleet and

its arsenal of nuclear bombs useless. At that time, the United States' primary deterrence to nuclear aggression from the Soviet Union was its vast air bomber fleets and vast number of nuclear bombs. Nuclear tipped missiles could easily destroy these vast armadas of bombers while they were still on the ground, in their hangers, or on the runway.

Second, it showed the United States to be behind the USSR in the race to control outer space. Not only could these missiles be equipped with nuclear warheads, but instead of placing a Sputnik type satellite into orbit, why not place nuclear warheads into orbit just circling and circling the globe? - So that at the flick of a button, warheads could be rained down upon the unsuspecting citizens below. Space itself could now be turned into a weapon. This was simply unacceptable to the American public.

Third, it demonstrated that the Soviet Union was ahead of the United States in space science and technology. Somehow the USSR was better at turning out trained scientists and engineers capable of creating such rockets and satellites. Prior to this, the United States saw itself as the superpower of the world. In one blow, Sputnik destroyed the illusion of the United States being the world's premier technological and scientific superpower. It showed that the USSR was not only equal to but likely superior to the United States in space technological and scientific abilities. This was not a position the people of the United States were accustomed to being in.

The coup de grace was given less than one month later when Sputnik II was launched by the Soviet Union. Sputnik II was much larger than its predecessor. It weighed an impressive 1,125 pounds and carried a dog named Laika into space. The satellite also contained a telemetry system, radio transmitters, a temperature control system, and a programming unit. Quite an amazing scientific achievement for 1958!

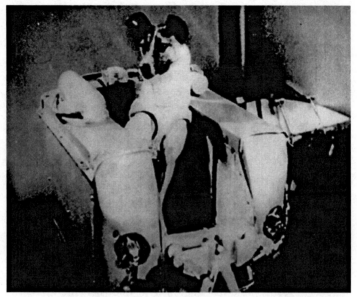

Figure 1.3: Sputnik II Liaka dog courtesy NASA

America was both scared and humiliated. Sputnik mania gripped the country. This was a humiliation and a threat the United States could not simply bear to let stand. Something had to be done. As one would expect, the United States government sprang into action.

First, within a year, the National Atmospheric and Space Administration (NASA) was created. This was done, if for no other reason, to get America's space program into high gear. This consolidated governmental efforts to get America into space. Prior to this, many government agencies were involved in the science of space exploration. NASA brought this all under one roof. This further provided more direct supervision of the entire effort by the one governmental agency, thus allowing for more rapid progress. In the end, the creation of NASA resulted in the Mercury, Gemini, and Apollo programs.

This eventually led to man's landing on the moon in July of 1969.

Second, the U. S. Army's Explorer Rocket program and the U. S. Navy's Vanguard missile program were put solidly on the front burner. Prior to this, they were just some of the many projects being completed by the Department of Defense. Now they were the projects being addressed by the Department. This led to a Vanguard satellite orbiting the earth on January 31, 1958. Unfortunately for the United States, this was a bit behind the Soviets.

Third, new education programs were created to better teach science and technology in America's public school system. The public school system in the United States had last seen such a major jolt to its core framework when Japan bombed Pearl Harbor. This was a revolutionary change. These new school programs included what became known as the "new math" and a much increased focus in all things science in the classrooms. Soon astronauts would become America's heroes, and cosmonauts would become heroes of the Soviet Union. The funding for these new programs in schools came from the National Defense Education Act (NDEA). Thanks to NDEA, funding for schools in the United States increased six fold by 1960 – or should I say, thanks, Sputnik.

Fourth, early in 1958, the Advanced Research Projects Agency (ARPA) was created. This agency was later renamed the Defense Advanced Research Projects Agency (DARPA). DARPA was created to perform research and development projects for the Department of Defense. Its goal was to prevent another surprise like that of Sputnik. It was to ensure the United States surprised its enemies, and was not surprised by those enemies. This agency's work led to the creation of ARPANET which in turn led to the creation of today's internet.

Fifth, the United States drastically increased funding to the National Science Foundation by a factor of close to four. The National Science Foundation is an independent federal agency tasked with among other things securing the national defense of the United States.

Sixth, with the election of President Kennedy in 1960, who campaigned heavily on closing the "missile gap" with the Soviet Union, thousands of new Minutemen Intercontinental Ballistic Missiles (ICBM) were built and deployed in the defense of the United States. These Minutemen Missiles were nuclear armed ballistic missiles hidden in silos across the United States. This "missile gap" was a direct outgrowth of the launch of Sputnik. The "missile gap" was a belief by the United States that the Soviet Union led the West in missile technology and in its deployment. This touched off an arms race between the two super powers that lasted for many decades and cost trillions and trillions of dollars and rubles.

Lastly, the Polaris Missile and Submarine Launched Ballistic Missile (SLBM) programs were fast tracked. These programs were given the highest priority in the United States. As pointed out previously, the United States government realized that if the Soviet Union could put a satellite into space, it could replace the satellite with an atomic warhead, thereby threatening America. Moreover, it realized that such a warhead would render the American strategic bomber fleet useless, as this fleet could be destroyed on the ground by such a warhead before the bombers could get airborne. This bomber fleet was the essence of the United States' strategic defense. Now it had been turned virtually useless by Sputnik. A new weapon had to be developed that could remain undetected until launch time. In that way, it would be invulnerable to attack by a nuclear missile. It is difficult to destroy what you cannot see and cannot find.

Figure 1.4: Polaris A1 Missile - courtesy U. S. Navy

Furthermore, the president as well as congress could also be destroyed by such a warhead, thus eliminating any retaliatory response. As one can see, citizens of the United States felt they had much to fear from Sputnik.

Lastly, it was envisioned that the USSR would need about five years to fully implement such a weapon. That meant that the United States had a short window of about five years to develop and deploy a weapon that would not be susceptible to destruction by an atomic weapon deployed from space. A weapon was needed that could remain virtually invisible and undetectable until it was needed to launch its complement of missiles.

Such a weapon did not exist at the time. Nothing even close existed. But one had to be created and quickly. A land based weapon would not do, as it could easily be eliminated by a nuclear strike. Land based weapons could not be made undetectable. This left only a sea based option. A sea or naval based option did not exist. Unfortunately, a surfaced naval vessel would equally not do, as it was also susceptible to attack from a nuclear missile or torpedo. Additionally, when a surface vessel is at sea, it is clearly visible. It is hard to hide in plain sight. This eliminated carriers, battleships, destroyers, transports, and cruisers as weapon platforms that could survive a nuclear attack. This left submarines as the only viable option.

Figure 1.5: U.S.S. George Washington SSBN 598 – the first U.S. Submarine capable of launching a nuclear missile while submerged. Courtesy U.S. Navy

Submarines would need to be created that could launch nuclear armed missiles while remaining submerged. This technology did not fully exist. Much of the necessary technology was still on the drawing board if it existed at all. A little progress had been made on such technology in the United States during the 1950s, but these projects were far from complete. Somehow this technology would have to be found, developed and deployed and in short order. A method would have to be developed that would guide this program to success. Project management and in particular the Program Evaluation and Review Technique (PERT) would become that method.

Scientific project management got its start when Henry Gantt created Gantt charts just prior to the United States involvement in World War I. Since then, little had changed in the project management world until the Polaris Project and the SSBN submarine program. Project management was revolutionized by Sputnik.

Polaris is a combination of a long range nuclear ballistic missile and a SSBN submarine. SSBN is the Navy's shorthand for a fleet ballistic submarine. This means it is a ship capable of launching nuclear ballistic missiles known as Submarine Launched Ballistic Missiles (SLBM). Polaris was the first of these SLBMs to be developed by any navy.

The Polaris Program and the SSBN Program were born of Sputnik. Sputnik forced a new reality on the United States and onto project management. The Polaris Program actually began in September 1955 but its progress before Sputnik was far from stellar. It was mired in interservice rivalry, making little to no progress. In February 1955, a United States National Security Council report recommended the creation of massive retaliatory response capability should the Soviet Union attack the United States. It also recommended a sea based option for the delivery of nuclear weapons to support this capability. From this report, two projects developed: Thor, Titan, and

Atlas by the United States Air Force and Jupiter by the United States Army and Navy. Neither project progressed very well nor would either project meet the requirements needed for an undetectable, submerged launching system.

After Sputnik, the Polaris Project took on a new sense of importance and purpose. From this, modern Project Management and modern risk management were born. As a direct result of Sputnik, Program Evaluation and Review Technique (PERT) was developed to ensure the Polaris Submarine program could be completed successfully and within the timeframe allowed. About the same time, the Critical Path Method (CPM) was developed by the DuPont Nemours Company, thus creating the foundation of modern project risk management.

This book intends to use the Polaris Submarine Project, which was the combination of the Polaris Missile Project and the SSBN Submarine Programs, as examples to demonstrate the proper use of modern project risk management techniques. Since these programs were only successful by the use of early risk management techniques, it only seems fitting they should be employed as examples to demonstrate modern project risk management. Many modern risk management techniques were not in existence when Polaris was completed. So naturally, Polaris could not have used them per se. Many of today's risk management techniques have their roots in the Polaris Project. For some risk examples used in the text, modern day processes are demonstrated using Polaris Project situations. But, quite frankly, for its day, Polaris was done right. Polaris is considered by many as the best project ever complete by the United States Military.

Unfortunately, information on many actual risks from the Polaris Project is not available, as much of the actual information on these programs remains classified for national security reasons. Many of the engineering and scientific discoveries of this project may still be in use today. Where details of actual risks are available, they have been incorporated into the text. Other risks were postulated based on the actual Polaris/SSBN Program and today's methods of risk management.

The Polaris Project created three missiles in a relatively short period of time. It created the A1, A2, and A3 designs. Each was an improvement over the last. This text will only examine the development and deployment of the first missile created: the A1.

## CHAPTER TWO

Risk Management Planning

Overall Process

Risk management, as with any area of project management, begins with planning. (See Figure 1, below.) One must decide how one plans on doing project risk management. There is no single best method to use for risk management, as one has many options from which to choose. The only wrong option is not to do any risk management, an all too common choice.

Figure 2.1: Overall Risk Management Process

As shown in Figure 2.1, Risk Identification follows Risk Management Planning. In Risk Identification, a number of tools and techniques are employed to identify risks associated with the project. Once these are identified, these risks are qualified in the next process. Qualitative Risk Analysis involves prioritizing the risks according to their significance to the project. Quantitative Risk Analysis is the next process in the series. This step provides numerous tools and techniques to further evaluate the risks to the project. Risk Response Planning creates response strategies for the most significant risks to the project. Risk Monitoring and Control monitors for new risks previously unidentified and ensures the response strategies created in Risk Response Planning are effective. A feedback loop joins Risk Monitoring and Control and Qualitative Risk Analysis to provide a mechanism to process any newly identified risks that are found after formal risk identification is complete. One must remember, risk management is an iterative process.

Risk Planning was critical for the *Polaris*/SSBN project, though, this was not necessarily a formal process at the time. It went by different names such as Systems Analysis, Operational Analysis, and Engineering Analysis. Risk management was not a very common term in those days. Much about risk management planning and project management in general was learned on the fly through this project and through the later Mercury, Gemini, and Apollo projects. Many of the techniques developed are still being used today to manage projects.

The Basics

To begin with the basics of risk management planning, it is best to define three terms: a *risk*, a *symptom*, and a *fact*. Many confuse these terms when speaking about risk management. This confusion can lead to disastrous results in managing projects.

A *risk,* according to the Project Management Institute's (PMI) *Project Management Body of Knowledge (PMBOK® Guide),* is defined as follows: "An uncertain event or condition that, if it occurs, has a positive or negative effect on a project's objectives."

A *symptom,* according to *Merriam-Webster Dictionary,* is "something that indicates the existence of something else".

A *fact,* according to *Merriam-Webster Dictionary,* is "something that has actual existence".

Many in project management list on the risk register (the risk register is a list of risks) going over budget or finishing behind schedule as risks. Neither one is a risk. From our definitions, they are symptoms or consequences of risk, but not risks. Later in this chapter, one will see they may be used as categories of risks, but they are not risks by themselves. If I attempt to treat them as risks, I would be treating the symptoms and likely not treating the disease. This does not work all that well in medicine, why would it work in project management? The risk is still out there not being corrected or treated, so the project continues to decline. One must find the underlying risks to be successful at risk management.

Let's look at another commonly listed risk on the risk register: resource unavailability. If I know at the beginning of a project that certain required resources are not going to be available when I need them, it is not a risk. It is a fact! That does not mean I disregard it. On the contrary, since resource unavailability is a fact, I do not need to decide if its rating/magnitude is high enough to bother with. Since it is a fact, I already know I have to do something proactive about this issue if it will have a significant effect on the project.

The overall point of risk management is to identify risks, to prioritize them according to a method of our choosing, to examine the risks with the highest magnitude, to consider and possibly implement actions to address the risks with the highest rating/magnitude, and to continually monitor the project for risks. The rating/magnitude of a risk is determined by its impact and its probability. There is little point doing this evaluation on facts, as their probability is already known to be 100%. Remember that 100% probability is not a probability at all; it is a fact! It will happen. Further, there is little point doing an evaluation of symptoms since they are not the underlying risks that are the cause of the symptom. I would end up just chasing my tail rather than managing risks. So, we need to focus on the real risks.

The Plan

The risk management plan should be the result of meetings with key stakeholders and fellow team members. It should not be the result of one person's effort. Risk management is too important to be left up to the designs of one person. Risk management planning for the *Polaris* Submarine Project involved a small core team and hundreds of support personnel. It certainly did not rely on one individual for such an important task.

Risk management planning begins with a review of the Organizational and Environmental assets of the organization. These include the processes and procedures one must follow while performing risk management, standard risk management plan formats, standard templates, industry and company-wide common definitions, the corporate culture, and lessons learned from past projects.

Most importantly, through this review of assets one must determine an organizational risk tolerance for this project. A risk tolerance is how much risk you are willing to accept to accomplish this project. Risk tolerance should vary by project. Certain projects are more important to the success of the corporation, thus perhaps adopting a higher risk tolerance may be preferred. Another project may be in an industry in which the possibility of something going wrong is unacceptable, thus a much, much lower risk tolerance may need to be adopted for those projects. Setting this risk tolerance is very important as it determines your risk threshold which is used in qualitative risk management. This threshold determines which risks should be addressed and which can be ignored based on the stakeholders' risk tolerance.

A risk management plan describes how risk management will be accomplished on our project. According to the *PMBOK® Guide*, a risk management plan should contain the following:

- methodology
- roles and responsibilities
- budget
- timing
- risk categories
- definitions of risk probability and impact
- probability and impact matrix
- revised stakeholders' tolerance
- reporting formats
- tracking

Each of these sections will be explained in more detail below.

Methodology

There are many choices of various methods to employ to perform risk management. In each section of the process, there are a number of possible methodologies from which to choose. In this section of the plan, we determine what methods we intend to employ later in the subsequent steps. Each method has its own required information needs. Also listed in this section of the plan will be how the data we need to perform the method(s) of our choosing will be obtained. The various methods available will be discussed in each method's respective section later in this book.

Roles and Responsibilities

Roles and responsibilities denote "who is responsible for what" with respect to risk management. Who will be the team leads for risk management? Who will support them? For each activity within risk management, a leader and support personnel will be established and listed here. Often a Responsibility Assignment Matrix (RAM) is created specifically for risk management. A RAM is often used in project management and typically matches project activities to those responsible for those activities.

Risk Management Activity	Danya	Danielle	Juan	Jill
Determine methodology	A	R		
Determine risk management budget and timing			A	R
Create RBS		A		C
Facilitate brainstorming session			A	
Perform SWOT		C		A
Perform force field analysis	A		C	
Do qualitative analysis	R	A	C	I
Perform Monte Carlo analysis			A	

Perform decision tree analysis		A		
Develop risk strategies	A	R	R	R
Assign risk owners	A	C	C	C
Monitor and control	A	R	R	R

Table 2.1: Sample Roles and Responsibility Matrix in RACI Format

In Table 2.1, roles and responsibilities are shown by the use of the letters R, A, C, and I. This is the RACI form of a RAM. The meaning of each letter is shown below in Table 2.2:

R	Responsible- this person is responsible for doing the activity
A	Accountable- this person insures the activity is completed
C	Consult- this person has direct input into this activity
I	Inform- this person is to be informed as to the work of this activity

Table 2.2: RACI Explanation

Using this method, we can clearly show who is responsible for performing the activity, who is accountable for the completion of the activity, who has an input into the activity, and who is to be informed of the work of the activity. In the activity "Perform Qualitative Analysis", Danielle is accountable for the completion of the activity, Danya is to perform the activity with input from Juan, and Danielle is to be informed when the activity is completed. This method clearly establishes responsibility and accountability for each activity.

A simpler form of a roles and responsibility matrix is also commonly used. This method simply lists the risk activities in one column and the name of the person responsible or group responsible in the other columns in the matrix. A simple X can be used to mark the square on the matrix where the activity and person/group responsible cross. It is important to create some sort of a role and responsibility matrix, as it assigns accountability to the activity. Without this accountability, critical risk management activities may not be completed on time or at all.

Budget

The budget section asks and answers the following question: How much do you plan to spend to do risk management? For a small project, this may be a very small amount of the total project budget. As projects get larger, the budget for risk management usually increases considerably. This budget should include costs of assigned resources, cost of methods to be employed, and cost of overall risk management. This section may also include how contingency reserves will be applied. Contingency reserves are funds set aside to help should known-unknown risks occur. A known-unknown risk is a risk that you have identified, thus it is known to exist, but you do not know if it will occur, thus whether it will actually occur is unknown. Often money is set aside for such risks in the form of a contingency reserve. We will discuss project reserves in greater detail later in this text. These risk management costs need to be included in the project cost baseline.

Timing

Timing, in Risk Management Planning, defines when and how often one will perform the processes associated with risk management. Essentially one is creating a schedule for risk management activities. This schedule should be included in the project schedule. While creating this schedule, one should consider how much time one plans on devoting to risk management. For a one week project, an hour may be plenty of time to do all risk management processes before the work begins and perhaps two more hours during the project to do monitoring and control. Obviously this would be far too little time for a six month or longer project. One must find a balance between the length and complexity of the project and how long one plans on expending time on risk management. Too much time spent may be a waste of time and money; too little time spent and the project may be plagued by risks.

Risk Categories

For ease of identification and later review, risks categories are normally created. These categories create a simple, systematic method to determine and evaluate risks. They create a common framework for viewing, managing, and controlling risks. Commonly these categories are shown in a Risk Breakdown Structure (RBS). This RBS is very similar to a Work Breakdown Structure (WBS) in that it breaks down risks into categories and subcategories much like a WBS breaks down work into deliverables, sub deliverables, and work packages. Risk Breakdown Structures should be tailored to each individual project. This is commonly accomplished by reviewing past projects and historical information and modifying this data to the current project.

Risk categories normally reflect the industry in which the project is taking place, thus many different categorization methods exist. Below are three general, non-industry

41

specific systems of categorization commonly employed in risk management.

Method 1	Method 2	Method 3
Quality Risks	Schedule Risks	IT Risks
Technical Risks	Quality Risks	Human Resource Skills
Project management risks	Cost Risks	Sales and Marketing Risks
Peformance Risks	Scope Risks	Operational Risks
External Risks	Resource Risks	Mmaintenace Risks
Organizational Risks		Financial Risks
		Communication Risks
		Engineering Risks

Table 2.3: An illustration of the results of different categorization options.

As one can see, a number of categorization systems exist. The key point to remember is that whatever categories you decide upon, the system must permit ease of communication and management of risk throughout the organization. You create the list based on the organization and one's project and you define each category on the list based on your individual needs. A RBS is really a tool to promote a mutual understanding of risks throughout the entire organization.

A number of methods also exist for displaying these categories. One method is to simply list the risk categories and sub categories. This is displayed in Table 2.4. A second method is to display the information in a more hierarchical format, as shown in Figure 2.2. Both show the exact same data but in different formats. Either format is fine as long as it promotes understanding and ease of management of the risks within your organization.

0.0 *Polaris* Submarine Project	1.0 Project Management Risks	1.1 Interface management
		1.2 Resources
		1.3 Support services
		1.4 Communications
		1.5 Controlling
		1.6 Planning
	2.0 Technical Risks	2.1 Requirements
		2.2 Security
		2.3 Performance
		2.4 Technology
		2.5 Processes
	3.0 External Risks	3.1 Political
		3.2 Legislation
		3.3 Foul weather
		3.4 Facilities
		3.5 External governments
		3.6 Contractors
		3.7 Suppliers
	4.0 Organizational Risks	4.1 Resource conflicts
		4.2 Prioritization
		4.3 Lack of support
		4.4 Dependencies

	5.0 Engineering Risks	5.1 Design
		5.2 Integrated testing
		5.3 Specialty engineering
		5.4 Feasibility of concept

Table 2.4: RBS shown in list format

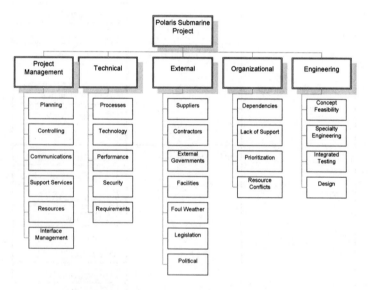

Figure 2.2: RBS shown in hierarchical format

For the *Polaris* Submarine Project, Figure 2.2 shows an appropriate RBS for the project. The *Polaris* Submarine Project was not your typical project. Many of the risks on this project were technical or project management in nature. Much of the work to be done had never been done before. There were questions as to whether the project was even possible. No one was really sure. To make matters worse, the team members as a whole were not project managers. They were engineers, technicians, and support personnel, but not project managers. External risks were also a major concern. Much of the work was to be accomplished by contractors. And, as with all governmental projects, governmental bureaucracy and politics were perceived early on to be major risks to the project. To that end, support for the project was questionable at best. Quite an inauspicious start, don't you think?

Definitions of Risk Probability and Impact

Risk management is all about probability and impact. Risk analysis revolves around these two concepts. But for them to mean anything we must define them for our current project. We must define what scales we will be using and what each level of our scale means before we define impact and probability. There are many different methods available to do this on our project. For our *Polaris* Submarine Project example, we will use a very low to very high scale for both impact and probability, as shown in Table 2.5. For each probability and impact scale, the figure shows the definition of each of the scales from very low to very high.

Probability	Schedule Impact	Cost Impact	Scope, Performance, or Quality Impact
Scale: Very High			
> 80%	> 6 months	> 10 million	Catastrophic reduction in original *Polaris*/SSBN program objectives. *Polaris* System unable to perform original function
Scale: High			
61-80%	2 months to 6 months	5 million to 10 million	Major reduction in original *Polaris*/SSBN program objectives
Scale: Medium			
41-60%	1 month to 2 months	1 million to 5 million	Moderate reduction in original *Polaris*/SSBN program objectives
Scale: Low			
21-40%	1 week to 1 month	250k – I million	Minor reduction in original *Polaris*/SSBN program objectives

Scale: Very Low			
1-20%	<1 week	<250K	Very minor impact to *Polaris*/SSBN program functionality

Table 2.5: Sample Risk Probability and Impact Scales

As one can see, these definitions and scales are very project specific. These scales should be created carefully, as they will form the basis for analyzing all project risks in the Qualitative Risk Analysis process. These create the common definition for communication concerning all risks. In other words, if I speak of a medium probability, high impact risk, I am speaking of a risk that has a probability of 41 to 60% and an impact of 5 to 10 million. I need not use the percentage or monetary amount; I simply call it a medium probability, high impact risk. It conveys the same information.

Other common scales exist. Instead of using very high to very low, one could use low-medium- high. Numerical scales are also employed in projects. These could be 1-3 corresponding to the low-medium-high scale or 1-5 scale corresponding to the very high to very low scale. Examples of each of these scales will be shown in much greater detail in Qualitative Risk Analysis in a later chapter.

Probability and Impact Matrix

Once definition scales are established, a matrix needs to be developed to utilize these definitions. After a risk is identified, this matrix along with the definition scales will be used to prioritize the risk in relation to all other risks. In Chapter Four, we will discuss various probability and impact matrices. Figure 2.3 is a sample risk matrix.

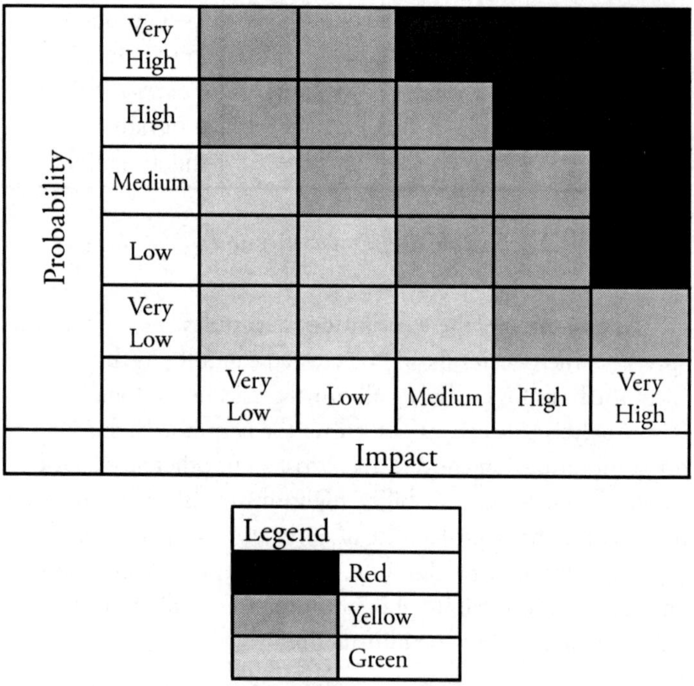

Probability		Very Low	Low	Medium	High	Very High
	Very High					
	High					
	Medium					
	Low					
	Very Low					
	Impact					

Legend	
	Red
	Yellow
	Green

Figure 2.3: Sample Risk Matrix

In Qualitative Risk Analysis (Chapter Four), one plots the probability versus the impact based on the previously created risk definitions. This will, as seen in Figure 2.3, color code or rank each risk. As you might expect, a red coded or ranked risk is much more significant to the success of the project than either a green or a yellow risk. Red risks are normally risks that require a response strategy of some kind. Orange is a fourth color found on some risk matrices. An orange colored risk, if used, is one that may be responded to if funds are left over after all the red risks are responded to appropriately. Yellow colored risks are generally watched over the course of the project. And a green colored risk is not necessarily addressed, as its impact and probability combination are too low for it to be of any significant risk to the project. In this way, we can prioritize each risk. We will discuss this process in more detail in a later chapter.

Your organization may have a standard matrix. If not, you may need to develop one. It is advisable to create a risk matrix that reflects the organization's view of risk for this particular project. See below for more on this subject.

Stakeholders' Tolerance

As stated earlier, understanding the risk tolerance for your project is critical to a project's success. Once this tolerance has been determined, it should be periodically reviewed to ensure that this tolerance still reflects the stakeholders' current view of risk on the project. Internal or external factors may have changed that require a new view of risk on a project. The Risk Matrix should track the current view of risk tolerance.

Stakeholders' tolerance of risk (or lack of tolerance of risk) should be reflected in the risk matrix. In Figure 2.3, we demonstrated a typical risk matrix. In Figures 2.4, 2.5, and 2.6, we will see how risk tolerance affects the risk matrix. We begin with Figure 2.4. It reflects a risk seeking approach. Notice only one block is in red and the addition of a new color: orange. Orange was added to reflect tolerance of very significant risks. This reflects the risk seeking approach. A risk seeking approach allows for the acceptance of more than normal risk on a project. Figure 2.5 shows a risk neutral tolerance. Risk neutral neither seeks nor avoids risk. It attempts to take a balanced approach to risk. In this approach, four blocks that were orange are now red signifying a less risk tolerant behavior, thus reflecting a risk neutral approach. Figure 2.6 shows a risk averse matrix. Risk aversion is a desire to avoid risk and uncertainty in projects. In this type of matrix nine out of 25 blocks are shown in red and the number of green blocks has dropped to four thus clearly demonstrating a risk averse tolerance.

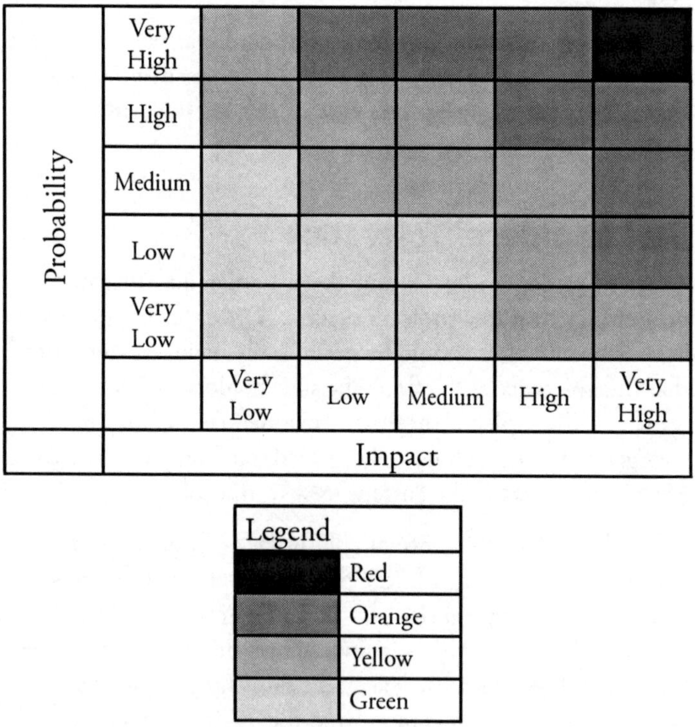

Probability						
	Very High					
	High					
	Medium					
	Low					
	Very Low					
		Very Low	Low	Medium	High	Very High
		Impact				

Legend	
	Red
	Orange
	Yellow
	Green

Figure 2.4: Sample Risk Seeking Risk Matrix

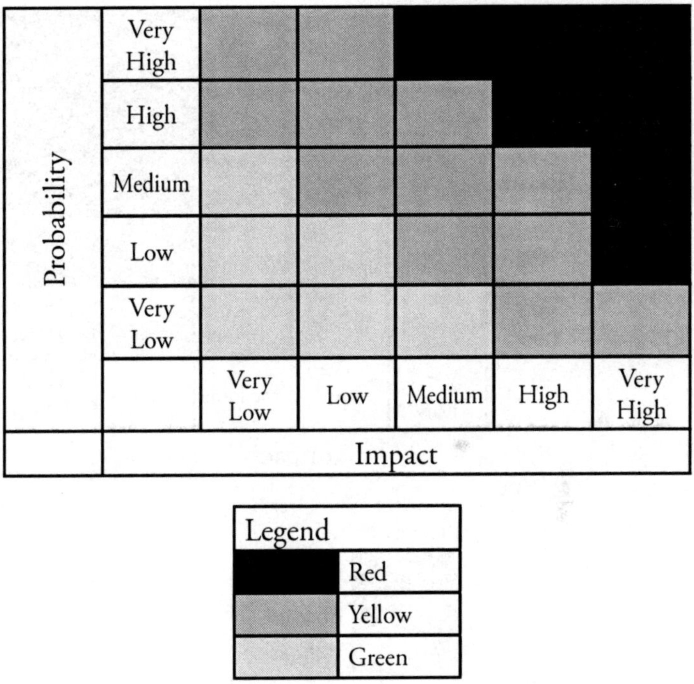

Figure 2.5: Sample Risk Neutral Risk Matrix

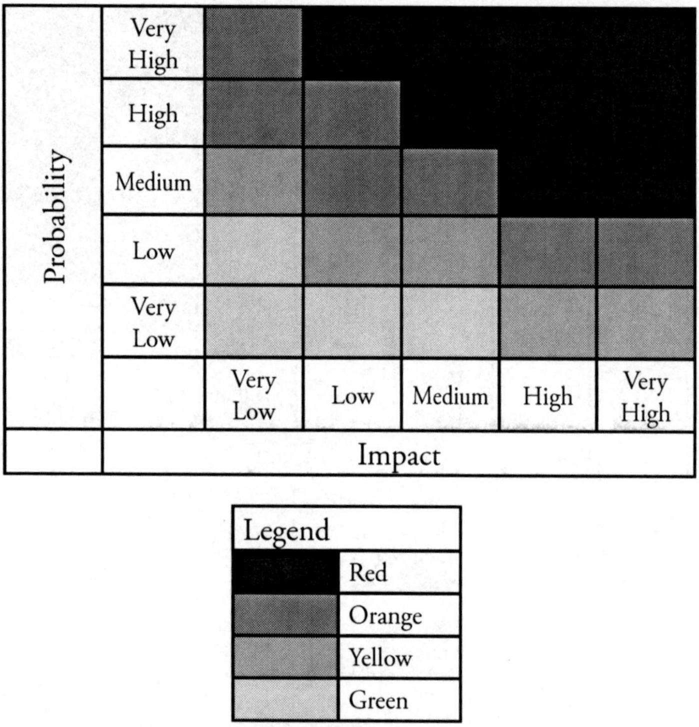

		Very Low	Low	Medium	High	Very High
Probability	Very High					
	High					
	Medium					
	Low					
	Very Low					
		Very Low	Low	Medium	High	Very High
	Impact					

Legend	
	Red
	Orange
	Yellow
	Green

Figure 2.6: Sample Risk Averse Risk Matrix

Depending on our stakeholders' tolerance, we need to develop a threshold for action. We see this threshold shown in the colors of the Figures. Assuming red is our threshold for action, a risk determined to be in the red area indicates a necessity for action of some sort with respect to this risk. An orange or yellow may indicate an action if the risk management budget permits. Green indicates placing the risk on a watch list. The watch list is a list of risks upon which no action will be taken at this time. These risks are watched and monitored during project execution. Chapter Four will discuss this in more detail. Thresholds are developed during this process for use in later risk management processes.

Let's examine stakeholder tolerance for the *Polaris* Project. To determine this, let's start by looking at what we did not have at the beginning of the project. We did not have a missile. We did not have guidance system. We did not have a propulsion system (liquid or solid fueled) that would work. We did not have a warhead for the missile. We did not have a reentry system for the warhead. We had no way to launch the missile from a submarine. From all or this, we probably should have a low risk tolerance for this project, as its possibility of success is very low. In that way, we would have reduced our losses if and when this project failed. **However**, if we could create a missile that could be launched from a submarine, a submarine that could remain undetected for weeks or months before launch, what then? How can you kill what cannot see? There had to be a value to that. What value is there in the possibility of such a system? If such a system **could** be developed, perhaps a more risk tolerant stance is appropriate. The possibility of actually creating such a weapon might outweigh the risks involved in attempting such a project. As such, the *Polaris* Project adopted a risk seeking tolerance versus a risk averse one.

Reporting format and tracking

Reporting format and tracking specify how and in what format documents associated with risk management will be completed. As an example, how will the risk register look? The risk register is the list of risks identified for this project. Much more will be discussed about the register in the next chapter. How will risk owners be notified and their progress tracked? A risk owner is an individual or group responsible for taking specific actions pertaining to a risk on the project. How will lessons learned be documented for use in future projects? How often, if at all, will the risk management process be audited during this project? Who will update the risk information on the project and how will this new information be communicated to the stakeholders? These are some of the questions to be answered in this section of the risk management plan.

## CHAPTER THREE

Risk Identification

Overview

Many risks existed that could have easily derailed the *Polaris* Project. Technical, engineering, and scientific risks abounded. These risks had to be evaluated and compensated for, if needed, before the project could be completed on time and on budget. This assumes, of course, that it could be completed at all. Given the circumstances, on time completion was an absolute must! But before any of this could happen, these risks had to be identified.

Figure 3.1: Risk Identification as part of the overall process

Many methods exist to identify risks. Some are better than others in certain circumstances. Each method has its pluses and minuses.

Before beginning risk identification, one must realize that both positive and negative risks exist. A positive risk is one that assists in achieving the project's goals and objectives. A negative risk does the opposite. Many individuals focus on the negative risks and do little with respect to the positive risks believing that positive risks either do not exist or are not to be controlled. Failing to consider positive risks may doom a project to failure.

For a project to be completed on time, each negative risk that occurs should be counteracted by a positive risk of equal magnitude. It is indeed a balancing act. One must consider and proactively work to enhance positive risks in order to offset negative risks that are undoubtedly going to occur.

Think of risks or opportunities this way: if this occurs (the probability), this will happen (the impact). Here is an example: If Russia abandons its space program, the president might abandon the United States' Space Program and thus the *Polaris* Project. This is a risk to our project. There was a probability that Russia might have abandoned its space program in the 1950s; though, this was a very low probability. They too had budgetary concerns to deal with. However, had this happened, the United States' President Eisenhower might have abandoned the United States space program and the *Polaris* Project. Unlikely, one would think, but possible. If this happened, there was a good chance this project would be canceled outright.

This possibility was actually much closer than many people realize. In March 1956, President Eisenhower refused to raise military expenditures to pay for such projects as *Polaris*. He felt that to do so would turn the United States into a "garrison state". For years, Eisenhower had been looking for a way out of the nuclear arms race. On December 8, 1953 Eisenhower made his famous "Atoms for Peace" speech to the United Nations in which he proposed contributing nuclear weapons material to peaceful atomic power programs – the Soviets did not respond. The President was also concerned about what effect continued high military expenditures would have on the economy. Further, in President Eisenhower's farewell address, he warned of America's growing military industrial complex and argued against it. Had President Kennedy followed President Eisenhower's lead, *Polaris* might never have succeeded. Instead, it would likely have been canceled or at least substantially reduced in scope. Happily for the project's future President, Kennedy campaigned on President Eisenhower's frugal military budget as a weakness and won the subsequent election. *Polaris* survived.

To begin risk identification, we begin by reviewing pertinent documentation associated with the project. As a minimum, review the project's scope, WBS, and any associated contracts. This documentation should inform you of what is to be done on the project, how this project is to be accomplished, any assumptions made concerning the project, and any associated plans for performing the work. A majority of the risks on a project will come from these areas. Another benefit of document reviews is the ability to review the quality of these documents. If it appears these documents were created in haste, that in and of itself is a risk.

The techniques below are used to review documents and the project as a whole to identify the underlying risks involved. The output of risk identification is a risk register. The risk register is a listing of risks with information about each risk. A risk register, when complete, should contain as a minimum the following:

- Project title and number
- Risk title
- Risk description
- Risk number
- WBS Number

- Risk owner
- Contingency plan
- Risk rating – probability and impact (Risk's effect on project objectives)
- Risk trigger
- Risk ranking
- Status: open/risk occurred/expired/closed/deleted
- Status date
- Risk response description
    o Actions taken to date and those planned
    o Was response effective?
- Notes

Risk registers are commonly created with spreadsheet software such as Microsoft Excel with the above as columns in a worksheet. Other projects have the risk register as a more formal document with each risk having its own page. This can be done in most word processing software such as Microsoft Word. In either event, at this point in time, much of this information to be inputted into the risk register database or into a risk register document is not available. Response/mitigation plans will not be created for some time yet. Information from subsequent steps within the risk management process will fill in missing data into the risk register. A sample risk register is found in Appendix B.

Risk Identification Methods

Brainstorming

Brainstorming is a technique in which participants are to solve a problem by generating a large number of ideas in a group setting. In this case, it is used to identify risks associated with a particular project or a particular WBS item within a project. Participants gather in a central location, ground rules are given, and participants are asked to identify risks. These ground rules usually entail not criticizing any idea presented and the welcoming of unusual, out of the box ideas. The desire is to obtain quality ideas to solve the problem presented. The technique is based upon the belief that, as the session progresses, ideas will build upon other ideas to create something more than the sum of its parts so that in the end 2+2=6, not four.

Participants are selected based on their expertise in a given field or phase of the project. A facilitator gathers the ideas presented in the session, enforces the ground rules, and may organize the ideas if desired. Surprisingly, smaller brainstorming groups seem to be more effective than large groups at producing more and better ideas.

Brainstorming is one of the most popular methods used in industry. However, there is no proof it is effective at producing ideas or for producing quality ideas. Evidence suggests that brainstorming is less effective than individuals working independently. In many cases, a small subgroup within the overall group dominates the discussion. This may cause what is known as groupthink.

Groupthink basically is a phenomenon in which the group subconsciously arrives at a consensus and reduces group conflict by not critically evaluating the issues at hand. This may occur when one individual or subgroup dominates the group work leading to a quick consensus. Groupthink is a major drawback in brainstorming and in decision making sessions, as these normally lead to a suboptimal group conclusion. It normally occurs when there is a strong, pervasive leader in the group and an intense pressure on the group for action. Groupthink has been linked to the Challenger Space Shuttle disaster and the Bay of Pigs invasion disaster.

Another drawback is that some group members do not participate at all. Some members may be intimidated by the group or, for whatever reason, do not feel inclined to participate in the brainstorming session. These drawbacks severely limit the effectiveness of the technique. Happily, brainstorming can be performed over the internet or by the use of software. This may reduce some of the effects of groupthink on the process.

From brainstorming, a number of risks were identified for the *Polaris* Submarine Project.

- Some scientists involved in the missile program may have ethical issues with the addition of nuclear warheads on the top of their missiles and thus may leave the program. This could cause some severe schedule risks.

- Can a missile be developed that is small enough to fit on a submarine and yet be able to fly 1500 miles as required by the project objectives? If not, this entire project may be at risk – a truly critical risk to the project!

- With all the recent (1950s) missile misfortunes, can missiles ever be tamed? If not, what can be accomplished with current technology? The American rocket and missile programs experienced a spate of missile failures during the 1950s and early 1960s. Every single one of the first 19 attempts by the United States to launch a rocket into space **failed!** These failures were casting doubt on America's ability to compete with the Soviet Union in missile technology. Some of these disasters were telecast on national television furthering the embarrassment and dismay felt by the American public. If missiles cannot be tamed quickly, it could create significant cost and schedule issues for the project.

- Can a Project Management methodology be developed that would assist in bringing this project in on time and on budget? None currently existed. If not, serious cost and schedule performance issues exist ahead.

Nominal Group

To reduce the adverse effects of groupthink in a brainstorming session, nominal group techniques may be used. Nominal group techniques are designed to engage the entire group in the brainstorming session. This is done by providing each participant with an equal say in the process. Instead of having participants voice their ideas verbally, as in brainstorming, in nominal group each writes their idea down anonymously. By writing down ideas anonymously, each and every participant is heard from and contributes equally. Even those that choose not to participate in a brainstorming session will be required to participate in a nominal group exercise. This increases the diversity of responses.

These ideas are then collected by the facilitator. The group then distills these ideas by voting on each idea presented without knowing who provided the idea. These ideas can be further distilled by subgroups if needed. In this manner, every participant has input into the process and has their individual ideas addressed.

Delphi

Delphi is another method that can be used to minimize the effects of groupthink. Delphi begins by developing a questionnaire designed to illicit the needed information. This questionnaire is then distributed to a group of subject matter experts. Responses to this questionnaire are collected and redistributed to the group with a second questionnaire. This method is performed anonymously, as the group members' identities are usually only known by the facilitator. Further, their responses are also distributed anonymously thus minimizing any effects from groupthink. This process is iterative and provides the participants more time to consider responses as compared to brainstorming or nominal group techniques. This method is excellent if the group of experts is not centrally located. Email, fax, and the Internet provide a simple method for performing the Delphi technique.

Unfortunately, the Delphi technique has been used in the past for unethical purposes. The Delphi technique has been used by governments, industry, and other institutions to force a predetermined result to occur. Once the questionnaires are collected, the results created by those individuals seeking to control the group output, are distributed instead of the actual responses from the experts. This is done to fool and mislead the participants. In the end, a specious outcome is the result. For this reason, many individuals are hesitant to participate in any Delphi sessions.

Interviewing Techniques

Interviewing involves performing interviews of experts, other project managers, the sponsor, and other stakeholders about likely project risks. Interview questions should be designed to elicit information concerning risks and should not be designed to acquire facts concerning the project. We are looking for risks, not facts. Hence these questions should be open ended to allow a free flow of information in the answer. Never use questions that can be answered with a yes/no or with the reciting of a simple fact. Little value can be obtained from such questions. Below are some examples of good and bad interviewing questions.

Good:

- What risks have you experienced in the past on similar projects?

- What problems do you see with the technical aspects of this work?

- What methods have you used on similar projects to reduce costs and shorten the project schedule?

- What can be done with the current project to increase its efficiency?

- What issues have you experienced with supplies and contractors in the past that are likely to be an issue with this project?

- What technical issues do you see affecting this project? Do you have any possible solutions?

Bad:

- How many projects have you worked on in the past?

- Do your projects usually run into problems?

- Is it possible to complete this project on time and on budget?

- Will there be technological problems with this project?

As one can see, the good questions attempt to extract information by using open ended questions. Open ended questions allow the responder to expound on the answer rather than just simply answering yes or no. A bad question does none of this. Instead, it asks for numerical or yes/no answers. Unfortunately, in practice, the yes/no or numerical answer seems to be asked more often than the open ended questions. A yes/no question provides little useful information. For example, "will there be technological problems with this project" if the answer is "yes", what valuable information did I obtain? Not really any useful data was obtained. I still do not know what the issues will be, when they are likely to occur, how likely are they to happen, and what the likely impact will be. All I do know is that there will be problems. I should have known that already.

Historical Records

Historical records should be reviewed from similar past projects. These records may include a wealth of information on risks associated with these projects. This process may be combined with interviewing of past project managers to obtain insight on their projects that may be useful in your project.

Actually, historical records existed that were of great benefit to the *Polaris* Submarine Project. The idea of launching missiles from submarines started many years earlier with the Germans in World War Two. The Germans actually fired a number of small artillery rockets from a submerged submarine, the U-511, in the May and June of 1942. Further, the Germans were working on a submarine that could fire a much larger V2 rocket at industrial centers in the United States. Thankfully, the war ended too soon for this method to be actually attempted against the United States. However, it is believed a German submarine rocket program was tested against Soviet harbor facilities late in the war. These German programs were beset from the start with technical issues. For one, the submarines became difficult to maneuver and control during launch. This could be a risk to the *Polaris* Project also. Another problem for the *Polaris* Program was that small artillery rockets used by the Germans were much, much smaller than the size of missile it would take to fly a nuclear warhead 1500 miles. But the Germans had shown that the basic concept did show promise.

SWOT

SWOT (Strengths, Weaknesses, Opportunities, and Threats) Analysis has been a technique used by industry for decades. But for application in project management risk identification, it is employed slightly differently. Instead of evaluating the Strengths, Weaknesses, Opportunities, and Threats to the project, we, in contrast, evaluate each with respect to the project team and overall corporation or organization.

Hence, we evaluate the Strengths of the project team and of the corporation/organization to determine project Opportunities – positive risks. We also evaluate the Weaknesses of the project team and overall organization to determine project threats – negative risks. In this manner, we examine risks not caused by what we are doing in the project or by how we decided to do the work, but, instead, we examine the human resource and organizationally caused risks to the project. This method will expose risks the other methods will likely miss.

Strengths	Weaknesses
• The *Polaris* project team contains some of the best scientific brains in the world • The United States Government is currently backing the project • The team members not only are some of the best theoretical scientists and engineers in world, they also have many years experience. Some getting their start in Germany during WWII.	• The *Polaris* project team has not entirely worked together before • Since the team members are so talented, some team members may leave the project to join non-governmental organizations for higher pay • The project team contains many highly educated members with little work experience • The team members are trained in physics, chemistry, engineering, and many other needed sciences, but not in project management
Opportunities	Threats
• Having the best theoretical knowledge and experience in the Western World may allow us to minimize the need for contract consultants thus reducing costs. • The backing of the United States Government may allow us the resources to crash many more activities than previously thought, thus shortening the project. (Crashing involves adding additional resources to an activity to shorten its duration.)	• The team lacks project management skills and they have not fully worked together. This may translate to poor management of the project. • By being educated in advanced sciences and engineering, the project team may not have the creativity needed to bring the project in on time and on budget. • The loss of talented team members to outside organizations may create significant resource issues that may reduce the team's ability to succeed on this project.

Table 3.1: Sample SWOT Analysis for the Polaris Submarine Program

As shown in the above example, the process begins by insightfully determining the project team and the organization's strengths and weaknesses. From these strengths and weakness, opportunities and threats are developed. These opportunities and threats are positive and negative risks to the project.

Another method for performing a SWOT Analysis on projects exists. In our example above, we looked at the strengths and weaknesses of both the team and the organization. We translated these into opportunities and threats. Another method examines the strengths of the team and separately the opportunities to the overall organization. Further, it considers weakness to the team only and threats derived from the overall organization. Both methods consider the same issues but in a slightly different light.

Assumption and constraint analysis

Assumptions are those things we believe to be true. Constraints are those things that restrict us or restrict our actions. They set up boundary conditions for our project. These should be listed in our project's scope statement. We base all our actions with respect to the project on these assumptions and constraints.

The problem is: what if they are not true and what if they should not have restricted us in the first place? Are we basing our project on beliefs that turn out to be false? Before waiting for that ugly truth to become a reality, why not ask ourselves that very question? But asking what if this is not true, we may expose a number of important risks to the project. These risks would probably have gone unnoticed until they occurred had we not asked these simple questions.

The same logic can be applied to constraints. Why not ask ourselves what if this constraint was removed. What other possibilities does that open up for us? We probably will not be able to remove most constraints imposed upon our projects, but if we are able to remove just one, we may be able to create something not considered previously.

So let's do an example of assumption analysis for the *Polaris* Submarine Project. We could start with the obvious assumption that this project was even possible. With all the technological obstacles that needed to be overcome, to some, even assuming the project was possible, might have been a stretch. But there was little value in that line of thinking from a risk management point of view. Instead, let's look at some of the more actionable assumptions. One rather large and looming assumption regarded the ability to design a system to guide the nuclear warhead to its target. For this project to be feasible, such a guidance system had to be created. Dropping an iron bomb from an airplane onto a target and missing the target is one thing, but firing a missile from thousands of miles away with a nuclear warhead attached and missing the target is another. Millions of lives could be at stake. Due to the enormous size of nuclear warheads in use at the time, much collateral damage was also possible. You had to hit the target you aimed at. Anything else would be disaster.

The problem was that no one was sure such a guidance system could be created. This is the crux of assumption analysis. One could not assume such a guidance system was possible given the technology of the day. What new risks did that create for the project? If we consider this question for a few seconds, many risks come to mind. Let's consider at a few of them below.

- If a reasonable guidance system is not possible, we may need to increase the size of the warhead to compensate for the lack of accuracy. This may require an increase in the size of the missile and the size of the missile tube, making submarine launch impossible. This was the current philosophy of the U.S. Air Force under General Curtis LeMay and his use of 10 megaton warheads, as he advocated larger and larger bombs to compensate for a lack of accuracy.

- If a reasonable guidance system is not possible, the project may need to be put on hold until such a system is possible. This could waste millions of dollars in expenses.

- If such a guidance system is not possible on a sea based system, then a land based option is the only short-term solution. This means that the *Polaris* Project is finished.

There are many more risks associated with this particular assumption that directly affect the project. Assumption analysis should create a whole new set of risks that need to be explored, as the ones above needed to be explored for the *Polaris* Project to ensure project success. This assumption analysis work was performed by the Naval Weapons Station China Lake, California. China Lake was tasked with performing a systems and operational analysis to examine each assumption associated with the *Polaris* Project.

Force Field Analysis

Force Field Analysis is another excellent tool for finding risks other methods may miss. Force Field Analysis was developed by Kurt Lewin for use with social sciences such as change management, social psychology, process management, and organizational development. This method is very much different from others described here. Instead of seeking risks associated with a project or a project methodology, it seeks risks caused by stakeholder forces for and against the project. In any project, there exist stakeholders driving for the project and stakeholders dead set against the project. In other words, you consider the external "need" for the project or lack thereof.

In Force Field Analysis, you begin by determining these individual forces. Next, you rate each force by such criteria as power and interest. Individuals with great power and great interest can be a source of either positive or negative risk depending on whether they are for or against the project.

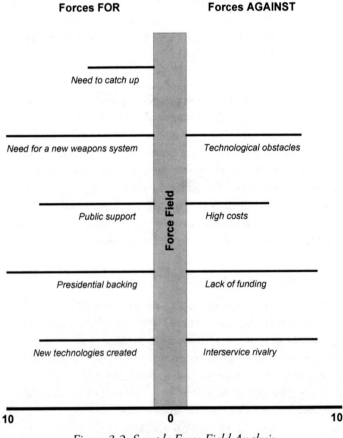

Figure 3.2: Sample Force Field Analysis

Either way, such individuals must be identified and planned for before a project can be successful. Other methods listed herein would likely not identify such risks. That makes this method essential in risk identification.

Figure 3.2 displays the forces for and the forces against the success of the project. Those for the project are called the driving forces and are shown on the left. On the right are displayed the forces against the success of the project; these are also known as the restraining forces. Each force is rated on how large of an influence it has on the success of the project. Driving forces are a need to catch up, a need for a new undetectable weapons system, public support, presidential backing, and that new technologies will be created. Each force is rated on a scale of 1 to 10. A one indicates little influence; a ten indicates a huge influence. The individual line length depicts the influence scale number. Presidential backing and the need for a new undetectable weapons system are each rated at level ten. This is the highest level a driving or restraining force can be. One can then total up the forces for the project. The total of the driving forces is 43, as shown below.

Force	Level
Need to catch up	5
Need for a new undetectable weapons system	10
Public support	9
Presidential backing	10
New technologies will be created	9
TOTAL	**43**

Table 3.2: Force Field Driving Forces

Restraining forces are technological obstacles, high costs, lack of funding, and interservice rivalry, as shown in Table 3.3. Lack of funding and interservice rivalry are the highest of the restraining forces in our analysis, as seen on our scale. Totaling the restraining forces we get 30 points against success. This is made up of the following:

Force	Level
Technological obstacles	7
High costs	5
Lack of funding	9
Interservice rivalry	9
TOTAL	**30**

Table 3.3: Force Field Restraining Forces

Looking at the total of the driving forces (43) and the total of the restraining forces (30) we see the project should be successful, as the driving forces are more significant on this project. However, this method does identify some major negative risks for the project: lack of funding and interservice rivalry. Just because the total for the driving forces is greater than that of the restraining forces does not mean we do not address these very influential negative risks. Something may need to be done to lower the level of at least these two risks.

The same can be said of our positive risks. We may decide to create responses for our opportunities also. Since presidential backing and a need of a new weapons system are both level ten, we cannot increase their effect. But we may need to do something to ensure they stay at level ten. The remaining opportunities are less than level ten, so they can be raised to a higher level if we choose to by creating a response plan.

Force Field analysis is a powerful tool to identify those risks caused by forces external to the project. Remember these risks are just as important to a project's success as any internal risk.

Checklist Analysis

Checklists are powerful tools to minimize the possibility of missing risks. Checklists are simple to construct. Past projects are useful guides in the development of these checklists. These checklists can simply list areas or categories in which risks need to be explored such as software, hardware, processes, procedures, technology, etc. They can be created as detailed as one needs to ensure as many risks as possible are identified. These checklists are normally based on the RBS created in the last chapter. Table 3.4 shows a sample risk checklist derived from the RBS created earlier. The checklist serves as simple method to ensure each category on the RBS is examined for its impact to the project.

Risk Category	Sub Category	Examined
1.0 Project Management Risks	1.1 Interface management	
	1.2 Resources	
	1.3 Support services	
	1.4 Communications	
	1.5 Controlling	
	1.6 Planning	
2.0 Technical Risks	2.1 Requirements	
	2.2 Security	
	2.3 Performance	
	2.4 Technology	
	2.5 Processes	

	3.1 Political	
	3.2 Legislation	
	3.3 Foul weather	
3.0 External Risks	3.4 Facilities	
	3.5 External governments	
	3.6 Contractors	
	3.7 Suppliers	
	4.1 Resource conflicts	
4.0 Organizational Risks	4.2 Prioritization	
	4.3 Lack of support	
	4.4 Dependencies	
	5.1 Design	
5.0 Engineering Risks	5.2 Integrated testing	
	5.3 Specialty engineering	
	5.4 Feasibility of concept	

Table 3.4: Sample Risk Checklist

Ishikawa Diagrams

The Ishikawa Diagram, fishbone diagram, or cause and effect diagram was originally created by Karoru Ishikawa as a quality tool for Kawasaki and has been used successfully by Mazda in the creation of the Miata. This diagram looks for relationships between the cause and effect of a given issue. In other words, this tool may be used to find the cause of various project issues. In that way, it is an excellent tool to postulate various problems or opportunities on the project in order to determine their possible causes – before they actually happen. Many companies use this method as their primary method of risk identification.

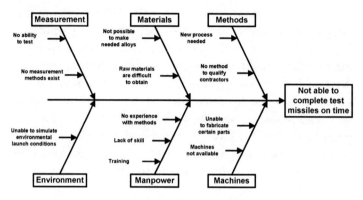

Figure 3.3: Sample Fishbone Diagram for Project

To begin using this technique, we begin with a problem statement. In the *Polaris* example above in Figure 3.3, the problem statement is that we are not able to complete the test missiles on time. In order to test our overall *Polaris* Missile System, we must have missiles available to use to test the system. The Fishbone Method can be used to determine the causes of events that have already happened, as well as to postulate possible events, such as the one in Figure 3.3. We are attempting to identify risks associated with events that actually have not happened in the project. For use in risk identification, we postulate events that may occur. Not being able to complete the test missiles on time would definitely put the project behind schedule and, hence, would be a significant event in the project. This would be a perfect candidate for risk identification.

Once a problem statement has been written, we next create major categories of possible causes. In our example, we chose measurement, materials, methods, environment, manpower, and machines. These categories may be particular for this specific project or can be generalized for all projects. They provide a framework for exploring risks.

Then, for each category, ask yourself, "With respect to this category, how or what could cause the problem statement?" For example, let's look at measurements. What issues with measurements could cause the missiles not to be ready on time? After deliberating on the subject, two issues come to mind. First, there may be no current method to test the missiles. So, how can I prove they actually work? Second, because this project will likely use very specialized manufacturing methods, measurement methods of individual components may not exist. These causal factors at this level are normally called primary causes.

Under each category, we see various issues on the above diagram associated with the problem statement. It is common to take this process down one more level to determine the contributing causes, or secondary causes, for each primary cause. In other words, what would have to happen for the primary cause to occur? This allows us to more precisely analyze the issue. As one can see, this method is an excellent method for use in risk identification. It is a good method to check the accuracy of other methods of risk identification used or to be used as a stand alone risk identification method.

Many of the risks identified above pertain to the ability to manufacture and test components used in the *Polaris* Missiles. These missiles were required to use as many "off the shelf" components as possible. This was done to speed construction. The military felt that using off the shelf components would allow faster construction of the missiles. The risk was that with such a poor success record – zero out of nineteen previously-attempted launches into orbit were successful – does not build trust in the ability to build such components and provide quality control over such components, making all the identified risks potential show stoppers.

Process Flow Charts

Process flow charts are easy to understand diagrams of the step-by-step process we create to do the project. By diagramming the process, risks not yet identified may become apparent. As they say, a picture is worth a thousand words. Once the process is visualized, new risks may jump out at you.

Diagramming begins by analyzing the process as it was delineated in the WBS and then building a step-by-step diagram that represents this process. The start and the end of the process are normally represented by elongated circles. These are connected to the steps or actions by lines. These steps or actions are shown as rectangles. Decision points within the process are shown as diamonds. Below is an example of a process flow diagram for the *Polaris/*SSBN Project.

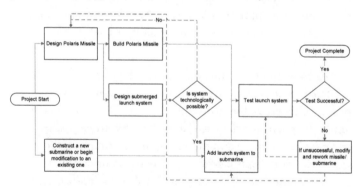

Figure 3.4: Sample Flow Chart for overall Project

Above is an appropriate flow chart for the entire overall *Polaris* SSBN Project. After the project starts, two independent processes begin. One is to design the actual missile. This process will involve obtaining funds, planning out the process, obtaining personnel and material resources, and doing the actual design and testing of this ground breaking missile. In order to get the project done in the required time, construction of the submarine will need to begin at the same time as design of the missile is occurring. This process, too, will require obtaining funds, gathering personnel and material resources, and actually constructing the boat.

Once design of the missile is complete (including all applicable testing), building of test missiles and designing a submerged launch system can begin. Building the missile will involve much contracting with vendors to build specific components of the missile and to assemble them into workable test missiles. A go/no-go decision point is placed after the design of the submerged launch system. This decision point represents the decision that will be needed in case it should become apparent that such a system is not technologically possible with the existing technological and manufacturing capabilities. If such a weapons system is not feasible at this time, this entire project may need to be reworked or canceled. As you will see later in this book, this did occur on the *Polaris* Project.

As shown in the figure, once a launch system is successfully created and the submarine construction is ready, the next process is to marry the two; in other words, to incorporate the newly designed launch system into the submarine. When the test missiles are ready and the submarine is complete, the next step is to test the entire system. If the test is successful, the project is complete. If not, Figure 3.4 shows a feedback loop to allow for any redesign, rework, or modifications needed to complete a successful test of the *Polaris* System.

This flow chart exposes one clear risk to the project: in order to complete the project on time, the submarine must be started before the feasibility of the missile launching system is proven. If a missile launching system is not possible, we could end up with a submarine that is essentially useless. This risk will need to be analyzed further in qualitative and possibly quantitative risk analysis. This risk to the project is too significant to ignore.

Influence Diagrams

Influence Diagrams are graphical depictions of elements and their relationships affecting a decision. The elements include chance variables, deterministic variables, decisions, and objects or results. Influence diagrams are designed to reveal the structure of a decision. They are not decision trees nor are they flow diagrams or flow charts. Decision trees will be explained later. Decision trees are usually more detailed and probabilistic in nature. As shown in Figure 3.5, flow charts and flow diagrams are sequential in nature and delineate a stepwise process. Table 3.5 provides a key to interpret influence diagrams.

◯ ⬭	Circles or ovals depict chance variables. These are variables that one cannot control.
▭	Rectangles depict decisions to be made.

	Double ovals or rectangles with rounded corners depict deterministic variables. These are variables that are calculated based on the variables feeding this node. If the deterministic variable was how far someone might travel in a period of time, then input chance variables to this node might be speed of travel and time traveling. These provide the information needed to calculate travel distance.
	Hexagons or diamonds represent the objective or results variable. This is the variable one is attempting to maximize or minimize.

Table 3.5: Influence Diagram Key

During the development of the *Polaris*/SSBN Program, it was decided to use solid fuel for the *Polaris* missiles. As you will see later, liquid fueled missiles simply would not work. The selection of any fuel for the missiles involves numerous considerations. Each one bringing with them their own risks. One risk associated with the choice of fuel is the risk of cancer from exposure to the fuel. Using an influence diagram, Figure 3.5, and the influence diagram key, Table 3.5, we will explore this important decision.

The Project team must decide on a propellant for the *Polaris* Missiles. That decision is represented by a rectangle. This decision will influence the cost of propellants and cancer related costs associated with the choice of propellant. The cost of propellants was actually influenced by the amount of energy a propellant could supply thereby limiting the range a missile could fly. Cancer related costs would be directly

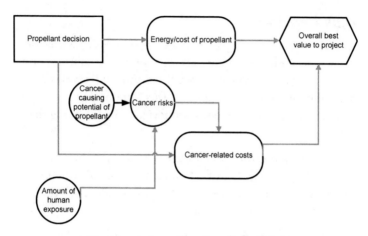

Figure 3.5: Sample Influence Diagram

related to the choice of propellant, as each propellant had its
own cancer causing properties. These two influence the overall
best value cost of propellant for the missiles. Cancer related
costs are a deterministic variable that can be calculated from
the overall cancer risk associated with the choice of propellant.
The overall cancer risk is a variable influenced by the cancer
causing potential of each propellant and the amount of human
exposure to the propellant. Both of these are shown as circles.
From this, we can see what influences the decision process for
this particular decision. Later in the process, as discussed in
Chapter Six, we can use this information to create risk response
strategies for this decision if this risk is significant enough to
warrant a response strategy. In the end, the energy of the
propellant was considered a more important influence in this
decision.

Risk Register

As stated earlier, a risk register can be as simple as a Microsoft Excel table with the required information in the columns or it can be a formal document. Table B.1 in Appendix B shows a more formal sample risk register. This form provides a simple method to document all needed information concerning the risk.

For our project, we will list the risks we have uncovered so far by using our risk identification techniques. We will list the information we have collected so far: risk number, risk title, and risk description. See Table 3.6, below.

One more important item may also be listed in the risk register at this point: risk triggers. A risk trigger is the symptom that a risk has occurred or is about to occur. This information may have been discovered during risk identification or may be discovered during subsequent steps. Risk triggers may be refined as we progress through risk management.

Risk #	Risk Title	Risk Description
1	Ethical issues	Some scientists and engineers involved in the missile program may have ethical issues with the addition of nuclear warheads on the top of their missiles and thus leave the program. This could cause some severe schedule risks.
2	Missile size	Can a missile be developed that is small enough to fit on a submarine and yet be able to fly 1500 miles as required? If not, this entire project may be at risk.

3	Inadequate missile technology	Is the state of missile technology advanced enough to allow this project to even be possible? If not, this poses a serious risk to the project.
4	Lack of a project management methodology	If a project management methodology cannot be developed to assist this project, it may not be possible to bring the project in on time or on budget.
5	Submarine control during launch	If a method to allow control and maneuverability of the submarine during launch cannot be developed easily, this may greatly extend the schedule of the project.
6	Expert knowledge	Having the best theoretical knowledge and experience in the Western world may allow us to minimize the need for contract consults thus reducing costs.
7	Government backing	The backing of the United States Government may allow us the resources to crash many more activities than previously thought, thus shortening the project.
8	Lack of project management skills	The team lacks project management skills, and they have not fully worked together. This may translate to poor management of the project.
9	Lack of creativity/ technical disputes	By being educated in advanced sciences, the project team may not have the creativity needed to bring the project in on time and on budget. This may lead to technical disputes among the project team.

10	Loss of talent to outside groups	The loss of talented team members to outside organizations may create significant resource issues that may reduce the team's ability to succeed on this project.
11	Guidance system	A reasonable guidance system for the missile may not be easily and quickly developed, thus causing schedule delays.
12	Lack of funding	President Eisenhower will not increase defense funding for this and other projects. If another source of funding is not secured, this project may have to cut scope and scale back the project.
13	Interservice rivalry	Interservice rivalry may reduce the current funding level to this project in order to support other projects. This will result in a major reduction in scope.
14	Need for the system	Since there is a need for an undetectable weapons system, an opportunity exists to increase funding to this project.
15	Presidential backing	If presidential backing can be maintained or increased, the possibility exists for increased funding to the project.
16	Public support	If public support can be increased then more pressure can be brought to bear on Congress to maintain or increase funding.
17	New technologies	The possibility of the creation of new technologies may permit an increase in support and funding for the project.

18	Testing of missiles	No easy method to test the missiles may exist. This may create schedule delays.
19	Measurement methods for components	Measurement methods for newly designed components may not exist resulting in schedule delays.
20	Simulate underwater launch conditions	Methods to simulate underwater launch conditions may need to be created. If these cannot be developed quickly and easily, cost and schedule overruns may develop.
21	New alloys	New alloys may need to be created for the missiles. If so, cost and schedule overruns are likely.
22	Difficult to get raw materials	It may be difficult to obtain raw materials needed for the project. This will result in cost and schedule delays.
23	New manufacturing methods	New manufacturing methods may be needed. If so schedule and costs delays are likely.
24	Contractor qualifications	New methods may need to be created to qualify contractors resulting in schedule delays.
25	Workers lacking skills	Workers may lack the necessary skills to fabricate new components. This may result in cost and schedule delays.
26	Workers lack training	Workers may lack the necessary training to fabricate new components. This may result in cost and schedule delays.
27	Workers lack experience	Workers may lack the necessary experience to fabricate new components. This may result in cost and schedule delays.

28	Components fabrication	Some required components may be difficult to manufacture resulting in cost and schedule delays.
29	Fabrication machines may not exist	Fabrication machines may have to be designed and developed in order to fabricate necessary components. This may result in both cost and schedule overruns.
30	Unusable submarine	In order to complete the project on time, the submarine must be started before the feasibility of the missile launching system is proven. If a missile launching system is not possible, we could end up with a submarine that is essentially useless.
31	Choice of fuel	The choice of fuel may increase the risk of cancer to individuals working on the project.

Table 3.6: Sample Risk Register from Risk Identification Process

## CHAPTER FOUR

Risk Qualification

R isk qualification and risk quantification form what is commonly known as risk analysis. Once risks are identified, they need to be analyzed. This process begins by having the risks prioritized. This is the crux of risk qualification. Believe it or not, all risks are not created equal. Some risks are more important, more dangerous, or, on the other hand, more desirable than others. Some risks if they occur will not impact a project much one way or the other. These can be put on a watch list and reviewed later. Some risks occur early in a project. So early in the project that they might occur before a formal evaluation is completed. This is another important concern. These risks many need to evaluated sooner rather than later. Other risks are so probable or so impactful that it is evident that something must be done vis-a-vis these risks.

One must start with the risk register created in the risk identification process. Each risk on this register needs to be

Figure 4.1: Risk Qualification as part of the overall process

evaluated based on its probability and its impact to the project. There are a number of methods available to accomplish this. The selection of method was made in the first process of risk management: risk management planning. For our purposes here, we shall explore quite a few possible methods.

Three Point Scales

Three Point Numeric Scales

A simple method is to rate each risk on a 1 - 3 scale for both impact and probability. This, like most of the methods in this chapter, gathers this 1 -3 point scale information from the project experts. The 1-3 point scale asks experts to rate each risk. A one means low impact if evaluating impact or low probability if discussing probability. A two on the scale indicates medium impact or probability, and a three indicates high impact or high probability. The impact and probability scores are multiplied together to get the overall risk rating. We see an example of this in Table 4.1 below. Our experts evaluated the first five risks on our risk register. Below are the results.

#	Risk	Proba-bility	Impact	Rating
1	**Ethical issues** Some scientists and engineers involved in the missile program may have ethical issues with the addition of nuclear warheads on the top of their missiles and thus leave the program. This could cause some severe schedule risks.	1	1	1

2	**Missile size** Can a missile be developed that is small enough to fit on a submarine and yet be able to fly 1500 miles as required? If not, this entire project may be at risk.	2	2	4
3	**Inadequate missile technology** Is the state of missile technology advanced enough to allow this project to even be possible? If not, this poses a serious risk to the project.	1	2	2
4	**Lack of a project management methodology** If a project management methodology cannot be developed to assist this project, it may not be possible to bring the project in on time or on budget.	1	3	3
5	**Submarine control during launch** If a method to allow control and maneuverability of the submarine during launch cannot be developed easily, this may greatly extend the schedule of the project.	1	3	3

Table 4.1: Sample Risk Rating on a 1-3 Scale

From the above Table 4.1, we see that after evaluating each risk and multiplying the impact and probability values together that risk number two is the highest rated risk. Risk numbers four and five are rated slightly less than risk number two. The other two risks, risks one and three, would be placed on the watch list. Whether any other of these risks would require responses depends on the threshold we created in risk management planning. If you remember the threshold was based on the stakeholder's tolerance of risk for this project. If the threshold for action was greater than a rating of four then none of these risks require response plans. Remember, to use this scale, we would have to use a three point scale for our Definitions of Impact and Probability from chapter two. In chapter two, we defined ours on a five point scale.

Three Point Color Scales

A second common method to prioritize risks is to substitute low-medium-high for the numbers 1-3 on our point scale. Using this method requires the use of a risk matrix as seen in Chapter Two.

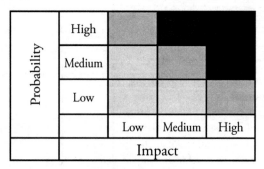

Figure 4.2: Sample risk matrix Low-Medium-High

The above matrix in Figure 4.2 shows a three point color scale. If our tolerance for action is red for any risk, then any risk that is evaluated as a high-high or high-medium or medium-high would require a response plan. The risks falling in the yellow zone may have response plans created if desired, and the risks in green should be placed on the watch list.

Five Point Scales

Five Point Numeric Scales

Another method to prioritize risks is the use of a 1-5 point scale. A one on this scale would indicate a very unlikely risk to occur or a risk with a very low impact should it occur. Of course, these are as defined in our Definitions of Probability and Impacts from chapter two: Risk Management Planning.

#	Risk	Proba-bility	Impact	Rating
6	**Expert Knowledge** Having the best theoretical knowledge and experience in the Western World may allow us to minimize the need for contract consults thus reducing costs.	3	3	9
7	**Government Backing** The backing of the United States Government may allow us the resources to crash many more activities than previously thought thus shortening the project.	4	5	20

8	**Lack of project managment skills** The team lacks project management skills, and they have not fully worked together. This may translate to poor management of the project.	4	4	16
9	**Lack of creativity / technical disputes** By being educated in advanced sciences, the project team may not have the creativity needed to bring the project in on time and on budget. This may lead to technical disputes among the project team.	5	5	25
10	**Loss of talent to outside groups** The loss of talented team members to outside organizations may create significant resource issues that may reduce the team's ability to succeed on this project.	1	3	3

Table 4.2: Sample Risk Rating on 1-5 Scale

From Table 4.2 above, we see risk number nine is of the most concern to the project. This is followed by risks numbered seven and eight. If our threshold for action is 20, then both risk nine and risk seven will need to have response plans created. Risk number eight would not necessarily require one. The remaining risks might be placed on the watch list.

Ease of Detection

An improvement on this technique adds one more number to the calculation: ease of detection. Ease of detection adds one more facet in the evaluation of the risk. The use of ease of detection in risk management is derived from Failure Mode and Effects Analysis (FMEA). FMEA is a technique to detect failure modes within a process or operation using similar past projects and then classifying these failure modes based on impact and probability. A more easily detected risk may be more easily and quickly responded to. A more difficult to detect risk may be much more difficult to respond to, as the risk may have been occurring for some time before detected. Ease of detection is also set to a 1 to 5 scale. A five indicates detection only after it has occurred for a negative risk or ample advance warning before if occurs for a positive risk. A one indicates detection with plenty of advanced warning for a negative risk or no advanced warning before occurring for a positive risk. Note positive and negative risks for ease of detection are opposites. In the below table 13, ease of detection is shown as EOD.

#	Risk	Proba-bility	Impact	EOD	Rating
6	**Expert Knowledge** Having the best theoretical knowledge and experience in the Western World may allow us to minimize the need for contract consults thus reducing costs.	3	3	5	45

7	**Government Backing** The backing of the United States Government may allow us the resources to crash many more activities than previously thought thus shortening the project.	4	5	4	80
8	**Lack of project managment skills** The team lacks project management skills, and they have not fully worked together. This may translate to poor management of the project.	4	4	5	80
9	**Lack of creativity / technical disputes** By being educated in advanced sciences, the project team may not have the creativity needed to bring the project in on time and on budget. This may lead to technical disputes among the project team.	5	5	5	125

10	**Loss of talent to outside groups** The loss of talented team members to outside organizations may create significant resource issues that may reduce the team's ability to succeed on this project.	1	3	3	9

Table 4.3: Risk Scales including Ease of Detection

As shown in the above Table 4.3, ease of detection clearly shows that risk nine is a very significant risk. However, with ease of detection in the mix, now risks seven and eight are equal with a rating of 80. Depending on our threshold, we may or may not create response plans for risks seven and eight.

Five Point Color Scales

A more commonly used method using a five point scale substitutes a color rating for a numerical one. To most people, a risk ranking of 25 may not mean a whole hill of beans, but a RED risk may. Colors invoke a preset understanding in people. We already know red is bad; green is good. Yellow and orange are somewhere in between. Five point color scales replace the 1-5 with very low-low-medium-high-very high scales. The results can be shown in three or four color scale. A four color risk matrix display is shown below in Table 4.4.

Probability	Very High					
	High					
	Medium					
	Low					
	Very Low					
		Very Low	Low	Medium	High	Very High
		Impact				

Legend	
	Red
	Orange
	Yellow
	Green

Table 4.4: Four Color Risk Matrix

In the above Table 4.4, risks are evaluated on a very low to a very high scale for both impact and probability. The results are plotted on the above matrix. This provides a color coded risk. Green risks are those whose combination of impact and probability show that they are not all that significant, so they can be placed on the watch list. The yellow risks are those whose risk rating shows they are more significant than a green risk, but due to budgetary restrictions, will most likely end up on the watch list. Orange and red risks are the critical risks for the success of the project. Red risks must be addressed. Their significance to the project is too high not to respond to. Orange risks may or may not be addressed depending on the project's threshold limit and the project's risk management budgetary limits.

Quality of Data

Qualitative and Quantitative Risk Analysis depend on data – good data. These data consist of probabilities, impacts, duration estimates, validity of the WBS, validity of resource estimates, validity of activity dependencies, validity of historical records, and many other important pieces of information needed to base our risk plans on. The question is – how good are these data? Is it just a shot in the dark or is it valid, dependable data? To do a reasonable Qualitative and Quantitative Risk Analysis, one must ask these questions.

Risk management requires accurate, unbiased data to be of any use. The less accurate the data, the less accurate the risk management is. It is key to verify the quality of the data one is using. If the data we are using is questionable and better data cannot be obtained, it is good practice to add buffers into response plans, budgets, and schedules.

Output

The output of risk qualification is a prioritized list of risks. This list takes the original list from risk identification and transforms it into a list of risks in order of their significance to the project. Those risks above the threshold will be forwarded on to Qualitative Risk Analysis, Chapter five, for further analysis. Those less than the threshold limit will most likely be placed on the watch list. Further evaluation of risks on the watch list is not required at this time. They will be "watched" to ensure their significance does not change over time.

It is important to remember that some risks may occur before a thorough analysis can occur. Risks that occur early in a project, before execution of the project begins are called urgent risks. These risks need to be evaluated before other risks, as they may occur before other risks.

Below in Table 4.5 is the updated risk register incorporating the risk rankings created using expert judgment. For this risk ranking, we will use a coloring coding of red, orange, yellow and green. In our ranking column in the example below, one will find not only the ranking but also whether the risk is a positive or a negative risk. Due to the threshold we created in risk planning the red risks will be forwarded to quantitative risk analysis. The remaining risks will be placed on the watch list.

#	Risk Title	Risk Description	Risk Ranking
1	Ethical issues	Some scientists and engineers involved in the missile program may have ethical issues with the addition of nuclear warheads on the top of their missiles and thus leave the program. This could cause some severe schedule risks.	Green Negative
2	Missile size	Can a missile be developed that is small enough to fit on a submarine and yet be able to fly 1500 miles as required? If not, this entire project may be at risk.	Red Negative

3	Inadequate missile technology	Is the state of missile technology advanced enough to allow this project to even be possible? If not, this poses a serious risk to the project.	Orange Negative
4	Lack of a project management methodology	If a project management methodology cannot be developed to assist this project, it may not be possible to bring the project in on time or on budget.	Red Negative
5	Submarine control during launch	If a method to allow control and maneuverability of the submarine during launch cannot be developed easily, this may greatly extend the schedule of the project.	Orange Negative
6	Expert knowledge	Having the best theoretical knowledge and experience in the Western world may allow us to minimize the need for contract consults thus reducing costs.	Red Positive

7	Government backing	The backing of the United States Government may allow us the resources to crash many more activities than previously thought, thus shortening the project.	Red Positive
8	Lack of project management skills	The team lacks project management skills, and they have not fully worked together. This may translate to poor management of the project.	Red Negative
9	Lack of creativity/ technical disputes	By being educated in advanced sciences, the project team may not have the creativity needed to bring the project in on time and on budget. This may lead to technical disputes among the project team.	Red Negative
10	Loss of talent to outside groups	The loss of talented team members to outside organizations may create significant resource issues that may reduce the team's ability to succeed on this project.	Yellow Negative

11	Guidance system	A reasonable guidance system for the missile may not be easily and quickly developed, thus causing schedule delays.	Red Negative
12	Lack of funding	President Eisenhower will not increase defense funding for this and other projects. If another source of funding is not secured, this project may have to cut scope and scale back the project.	Orange Negative
13	Interservice rivalry	Interservice rivalry may reduce the current funding level to this project in order to support other projects. This will result in a major reduction in scope.	Yellow Negative
14	Need for the system	Since there is a need for an undetectable weapons system, an opportunity exists to increase funding to this project.	Orange Negative
15	Presidential backing	If presidential backing can be maintained or increased, the possibility exists for increased funding to the project.	Red Negative

16	Public support	If public support can be increased then more pressure can be brought to bear on Congress to maintain or increase funding.	Yellow Positive
17	New technologies	The possibility of the creation of new technologies may permit an increase in support and funding for the project.	Orange Positive
18	Testing of missiles	No easy method to test the missiles may exist. This may create schedule delays.	Orange Negative
19	Measurement methods for components	Measurement methods for newly designed components may not exist resulting in schedule delays.	Yellow Negative
20	Simulate underwater launch conditions	Methods to simulate underwater launch conditions may need to be created. If these cannot be developed quickly and easily, cost and schedule overruns may develop.	Yellow Negative
21	New alloys	New alloys may need to be created for the missiles. If so, cost and schedule overruns are likely.	Yellow Negative

22	Difficult to get raw materials	It may be difficult to obtain raw materials needed for the project. This will result in cost and schedule delays.	Green Negative
23	New manufacturing methods	New manufacturing methods may be needed. If so schedule and costs delays are likely.	Green Negative
24	Contractor qualifications	New methods may need to be created to qualify contractors resulting in schedule delays.	Yellow Negative
25	Workers lacking skills	Workers may lack the necessary skills to fabricate new components. This may result in cost and schedule delays.	Yellow Negative
26	Workers lack training	Workers may lack the necessary training to fabricate new components. This may result in cost and schedule delays.	Yellow Negative
27	Workers lack experience	Workers may lack the necessary experience to fabricate new components. This may result in cost and schedule delays.	Green Negative

28	Components fabrication	Some required components may be difficult to manufacture resulting in cost and schedule delays.	Green Negative
29	Fabrication machines may not exist	Fabrication machines may have to be designed and developed in order to fabricate necessary components. This may result in both cost and schedule overruns.	Green Negative
30	Unusable submarine	In order to complete the project on time, the submarine must be started before the feasibility of the missile launching system is proven. If a missile launching system is not possible, we could end up with a submarine that is essentially useless.	Yellow Negative
31	Choice of fuel	The choice of fuel may increase the risk of cancer to individuals working on the project.	Yellow Negative

Table 4.5: Sample Risk Register after Qualitative Risk Analysis

After reviewing Table 4.5, it is clear some risks are much more significant to the project than others. Assuming red risks are at or above our threshold level, the following risks require further evaluation in the next chapter, Quantitative Risk Analysis:

#	Risk Title	Risk Description	Risk Ranking
2	Missile size	Can a missile be developed that is small enough to fit on a submarine and yet be able to fly 1500 miles as required? If not, this entire project may be at risk.	Red Negative
4	Lack of a project management methodology	If a project management methodology cannot be developed to assist this project, it may not be possible to bring the project in on time or on budget.	Red Negative
6	Expert knowledge	Having the best theoretical knowledge and experience in the Western world may allow us to minimize the need for contract consults thus reducing costs.	Red Positive

7	Government backing	The backing of the United States Government may allow us the resources to crash many more activities than previously thought, thus shortening the project.	Red Positive
8	Lack of project management skills	The team lacks project management skills, and they have not fully worked together. This may translate to poor management of the project.	Red Negative
9	Lack of creativity/ technical disputes	By being educated in advanced sciences, the project team may not have the creativity needed to bring the project in on time and on budget. This may lead to technical disputes among the project team.	Red Negative
11	Guidance system	A reasonable guidance system for the missile may not be easily and quickly developed, thus causing schedule delays.	Red Negative

15	Presidential backing	If presidential backing can be maintained or increased, the possibility exists for increased funding to the project.	Red Negative

Table 4.6: Risks Requiring Further Analysis

Notice on Table 4.6 that some of the risks are positive and some of the risks are negative. Red does not always mean bad. The color red for a positive risk is a good thing. It indicates an opportunity that should be explored. It is an opportunity to possibility shorten the project schedule, reduce budget, or improve the quality of the project results. Red for a negative risk is a risk that threatens the project's schedule, budget, or performance. These may need to be addressed if they prove significant to the outcome of the project. These red risks will be further evaluated in next chapter, Quantitative Risk Analysis.

Quantitative Risk Analysis

Quantitative risk analysis performs many functions in risk management. It further evaluates the risks that exceeded the threshold in qualitative risk analysis by using Expected Monetary Value (EMV). It develops and utilizes simulations and models to further evaluate the project. It can be used to calculate a project's standard deviation and variance that can be used to compare the riskiness of various projects. It evaluates decisions important to the project in an objective manner versus a subjective manner. Lastly, it can be used to calculate the contingency reserve for the project.

Figure 5.1: Quantitative Risk Analysis as part of the overall process

The primary methods used to perform these functions are Program Evaluation and Review Technique (PERT), Monte Carlo Simulations, Expected Monetary Value (EMV), Decision Tree Analysis, and Contingency Reserve/Reserve Analysis.

PERT

PERT was the method developed to manage the *Polaris* Submarine Project successfully. It is a method designed to determine the probability of completing a project on time based on the risks associated with the project. PERT is a series of calculations. Below is an example of a PERT calculation.

PERT begins by using expert judgment to determine three possible durations for each activity on the project. An expert or, if possible, a group of experts are asked to provide an optimistic, a pessimistic, and a most likely time for each activity on the project. This is known as a three point estimate. Most projects use only one duration estimate for a given activity. This single estimate is normally known as the most likely estimate. It has been shown that by using a three point estimate, the estimate of the true duration is significantly improved. For our project, we will base our time estimates on the *Polaris* flow chart below.

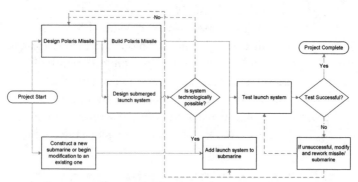

Figure 5.2: PERT Example Flow Diagram

For each of the process above a three point estimate is created. These estimates are shown in Table 5.1. A time estimate (TE) is calculated from this three point estimate by using the equation shown below. All values are shown in months for this project. As one can see, PERT uses a weighted average equation. This equation puts more weight on the most likely duration and less on the optimistic and pessimistic durations. This usually leads to a calculated TE that is more closely aligned with the actual duration of the activity.

$$TE = \frac{\text{Optimistic} + \text{Pessimistic} + (4 \times \text{Most Likely})}{6}$$

Process	Opti-mistic	Most Likely	Pessi-mistic	TE
Design Polaris Missile	18	24	30	24
Design submerged launch system	4	7	11	7.2
Construct/modify submarine	8	24	36	23.3
Build Polaris missile	4	8	12	8
Add launch system to submarine	6	8	10	8
Test launch system	1	2	5	2.3

Table 5.1: TE Values from PERT

A project's critical path is that sequence of activities from start to finish that determines the overall project duration. If a critical path activity takes one day longer to finish, then the project must take one day longer to finish; hence, critical path activities are the most important to the project's success. We add the time estimate (TE) values to the flow chart and get a critical path for the *Polaris* Submarine Project of the following: design *Polaris* Missile, design launch system, add launch system to submarine, and test launch system. This leads to a project duration of 41.5 months. For the critical path processes, we now need to calculate a standard deviation and a variance.

Non critical processes/activities do not contribute to the critical path, and, therefore, can be disregarded for this calculation

A Standard Deviation is calculated by the equation:

$$\sigma = \frac{\text{Pessimistic} - \text{optimistic}}{6}$$

A variance is the standard deviation^2 or σ^2

Process	Standard Deviation (σ)	Variance
Design Polaris missile	2	4
Design launch system	1.16	1.3456
Add launch system to submarine	.67	.4489
Test launch system	.67	.4489

Table 5.2: Standard Deviations and Variances – PERT

This table tells us a lot of useful information. The closer to zero a standard deviation calculates out to - the better for the project. This means less risk. Standard deviations above approximately 1.8 indicate much more risk in a project. Hence, "add launch system" and "test launch system" are low risk processes. As seen by its Standard Deviation, Designing *Polaris* Missile is a very high risk process. If we were to focus on any given process to reduce risk in the project, we should focus on designing the *Polaris* Missile. Designing the launch system is next in line, as one can see by its high standard deviation and variance. This can easily be seen on a tornado diagram, which will be described later in the Monte Carlo section.

The next step in the PERT process is to add all the variances on the critical path together. This leads to a result of 6.2434. Since we have been using the square of the standard deviation to get the variance, we now need to take the square root of this number to get back to a standard deviation. This calculates to approximately 2.5. This number is the standard deviation of the entire project. This is a very important number! This number can be compared to the standard deviations of other projects to see the relative riskiness of various projects if any such similar projects exist. A standard deviation of 2.5 is considered a very risky project. Again, the closer the number is to zero - the better for the project. A standard deviation range is approximately 0 to 3. As one can see, a result of 2.5 is very high on this range. This indicates the *Polaris* Project to be a high risk project. A large part of the risk on this project comes from designing a *Polaris* missile. This can be seen by the variance of 4 and a standard deviation of 2 in Table 18.

To enhance the likelihood of success of the *Polaris* Project, one must address the risks associated with the process Design *Polaris* Missile. One may also have to address the risk Design Launch System. The individual risks associated with each process are shown in the risk register.

PERT is not finished here. PERT will also calculate the probability of completing this project on a given date. This is another key purpose of PERT. To do this, we need only the numbers we already calculated, a given finish date, and a probability table.

The Intermediate Range Ballistic Missile program actually began in 1955-1956. This is the program *Polaris* was derived from. The name *Polaris* actually came later in the program. The Program had only made marginal progress before the launch of *Sputnik*. The United States was working on a few different missile programs at this time. As mentioned earlier, the United States Navy was working on the *Vanguard* Missile Program. The United States Army and Navy were also working jointly on the *Jupiter* Missile Program to name a few of the programs in progress. The *Jupiter* Program was a cruise missile program that required a submarine to surface before launching. The United States' missile programs at this point in time were a diverse group with little common purpose. Each branch of the service had its own budgetary and importance concerns associated with its missile programs. The *Jupiter* Missile was found to be too large to fit easily on ship and was liquid fueled making it impractical for shipboard use. Hence, the United States Navy withdrew from the *Jupiter* Program in 1956 to focus on solely the *Polaris* Project. At the time, governmental bureaucracy and interservice rivalry were stifling the *Polaris* Project. *Sputnik* created a single sense of purpose for the missile programs of the United States. Further, none of these missiles in development met the requirements set forth for *Polaris*. Cruise missiles in development at the time lacked the accuracy required for nuclear warheads.

There is an old adage that you only need to be close with horseshoes, hand grenades and nuclear weapons. That is so untrue. For hand grenades it may be true, but not for nuclear weapons. The *Polaris* weapon needed to fly at least 1500 miles and hit a target with a nuclear weapon. Later generation missiles would be able to fly half the world before hitting a target with nuclear weapons – close is not good enough; you must hit the target! Missing at half the world means you miss by entire countries, not cities.

It was imperative for the United States to ensure that this program was completed on time or completed early. And, the missile had to be able to hit the target. The security of the United States and that of the free world depended on it. Some way needed to be created to determine not just what activities/processes had high risks, but whether this project had a reasonable chance to be completed on time. PERT can do that calculation.

As shown earlier, the project began in 1955-1956, but it really got put into overdrive after *Sputnik*. The *Sputnik* launch was in late 1957, so let's use 1958 for the actual start of this project. Using PERT, let's determine the probability is of completing the project by the beginning of 1962. We need to use another equation to do that. This equation is shown below.

$Z = \dfrac{D - S}{\sigma}$
D= desired end time
S=calculated end time
σ=Standard Deviation of the project

Table 5.3: Determining the probability of on-time project completion.

So for our project, D would be 36 months, since it is 36 months between 1958 and the beginning of 1962. The calculated end time was 41.5 months as shown above. And the standard deviation of the project was 2.5.

Hence, Z= (36-41.5)/2.5 or -2.2. Using standard Z score tables, found in Appendix C, we see a score of .9861. Z score tables are read by finding the 2.2 on the left hand column and reading across to find the intersection with the column showing the second decimal place. This intersection is the Z score. Next, because our Z score is a negative number, we must subtract 1 from the number we read off the table in Appendix C. This leaves .0139 or 01.39% after multiplying by 100 to compensate for the percentage.

This is a less than two percent chance of completing the project by 1962 as risks stand now in the project. This number shows a number of important things. For one, it shows the current risk status in the project and shows that it is virtually impossible to complete by 1962. Second, it shows which processes/activities need to be addressed to provide a better chance of completing the project on time. We saw this by the processes/activities with the higher standard variations/variances. Lastly, it shows the value of PERT. Before PERT, no method existed that could provide us with this information.

Let's do another example. Let's see what the probability of success would be if we targeted 1963 as the completion date. Again we need to calculate a Z score. In this case, 48 would be the desired end time or our D number. All other numbers would be the same. So our equation would be as follows:

$$Z = (48-41.5)/2.5.$$

This leads to a Z score of 2.6. Since this is a positive number we can go directly to the Z score chart in Appendix C and not be concerned about subtracting the result from one. The Z score directly from the chart is .9953. After multiplying by 100 to translate this number into a percentage, we get 99.53%. This tells us fairly conclusively that the project should be completed by the beginning of 1963.

This is as the project stands now. The project will end somewhere between 1962 and 1963. Since we know this date is unacceptable, we must create responses to our most significant risks. This will be done in the risk response planning chapter.

Monte Carlo Simulation

Monte Carlo simulation is like PERT on steroids. It does all that PERT does and much more. Monte Carlo simulation, as it has evolved, is as close to a "do all" technique in risk management as there is. Much of PERT was designed to be accomplished using a slide rule or a rudimentary computer. Monte Carlo, as it exists today, is designed to use modern computers in conjunction with risk management techniques to efficiently and effectively manage risks.

When I first learned the Monte Carlo technique, I was told the simulation was originally a simulation used to predict gambling results; hence, it was named after the fact that casinos are a commonplace feature in Monte Carlo. I found out later that this is a myth. Later, I was told the simulation was created and designed to predict the landing point of nuclear artillery shells. This a little closer to the truth, but I would think that no matter where a nuclear tipped artillery shell lands, it is going to affect the gun that fired it as well as the impact point. Maybe not to the same degree, but both will be affected. This can be predicted without any special technique. In truth, Monte Carlo was actually developed for the Manhattan Project: the project to create the atomic bomb. It was the brainchild of Enrico Fermi in the 1930s. It was designed to predict various probabilities associated with the creation of the atomic bomb.

Monte Carlo simulation has progressed much since then. Monte Carlo was originally limited by the computer technology available at the time. As computing advanced so did the study of Monte Carlo simulation. Today's Monte Carlo simulation is available in software packages available for all project managers. For the most part, Monte Carlo software performs most, if not all, typical functions needed for risk management and project scheduling on a project. Below is a list of some of the features of today's Monte Carlo software. These features vary from product to product.

- Full schedule and resources engine. It will do everything typical project management software will do

- Maintains a full risk register directly linkable to the project schedule

- Qualitative and quantitative analysis of the project

- Probabilistic cash flow analysis

- Project scheduling and monitoring incorporating identified risks

- Decision tree analysis

- Analysis of risk responses

- Tornado diagrams to determine most significant risks

- Sensitivity analysis

Below is a typical risk register found in Monte Carlo software. With most of today's Monte Carlo software, risks are

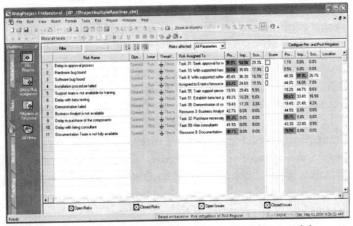

Figure 5.4: Monte Carlo Risk Register (Courtesy of the Intaver Institute)

tied directly to a given activity or the entire project itself. In this way, the effects of the risk can be directly viewed on that activity and the project as a whole.

From this probabilistic information, a schedule and budget can be created. This is shown in Figure 5.5. Notice in the figure the three point duration estimates. If you remember from PERT, these were based on expert opinion as to the optimistic, pessimistic, and most likely times. In Monte Carlo simulation, these are based on the effects of individual risks associated with a given activity. At the time PERT was created, computing power was not sufficient to do such calculations easily.

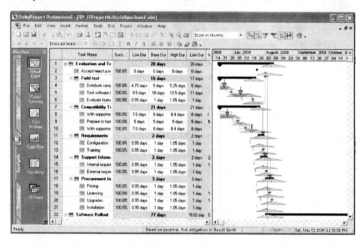

Figure 5.5: Monte Carlo Based Schedule with risks included (Courtesy of the Intaver Institute)

Based on the input risk data, a risk based probabilistic schedule and budget are created to predict a finish date and ending budget amount. This is shown in Figure 5.6. This budgetary information can be further broken down to predict cash flow on a periodic basis for the entire project.

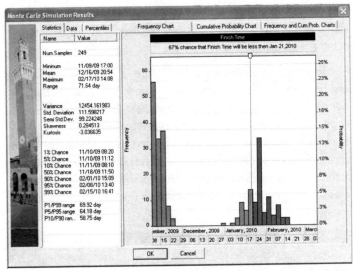

Figure 5.6: Monte Carlo End Date Calculation (Courtesy of the Intaver Institute)

If this end date of the project is unsatisfactory, Monte Carlo software packages provide Tornado Diagrams of the risks to assist in determining which risks require response plans to improve the simulation's results. Most Tornado Diagrams are based on a qualitative or quantitative calculation for each individual risk, activity, or task. A tornado diagram is shown in Figure 23. This normally is called "Sensitivity Analysis."

Sensitivity analysis measures how much a change in one parameter will affect the project, such as the project's duration or schedule. In other words, if a given risk were to occur, how much would that affect the project's duration or schedule? In this way, we see which risks provide the biggest bang for our buck when it comes to deciding which risks should have response plans created and which ones should not.

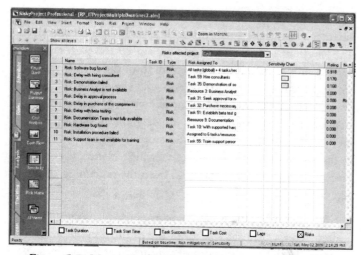

Figure 5.7: Monte Carlo Tornado Diagram (Courtesy of the Intaver Institute)

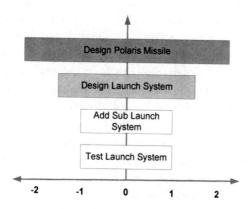

Figure 5.8: Sample Tornado Diagram for the Polaris Project

The information in Figure 5.8 was created in the previous section of this chapter on PERT, in Table 18, when standard deviations were calculated for the Polaris project. For this example, the X axis displays standard deviations on a scale of

121

-2 to 2. Other values can also be used for the X axis such as variance. Along the Y axis our critical path tasks are displayed. The length of each displayed box for our critical path tasks corresponds to its standard deviation from the X axis. Tasks are displayed on a tornado diagram with the higher standard deviations tasks at the top and are shown with decreasing standard deviation values down the diagram. Using this method, we can easily see that the more significant risks rise to the top of the chart for ease of viewing. Hence, the critical project risks or activities standout from the remaining risks or activities.

Monte Carlo can also evaluate response plans created to address significant risks and rerun the simulation- all within seconds! In this way, one can determine what if any affect the created response plans will have on the project as a whole.

One last note on Monte Carlo software, some Monte Carlo software only performs financial risk analysis, not project risk analysis. Most Monte Carlo software does financial risk calculations such as Net Present Value, Pay Back, Discounted Pay Back, Profitability Index, and Internal Rate of Return. These are critical in determining whether to perform a project or not. Unfortunately, many software packages stop there. Such software would be of limited value to a project manager since project managers are normally assigned after the project has been given the authorization to proceed. For a version of Monte Carlo to be of value to project managers or a project risk manager, it must do project risk management. Be careful when you select Monte Carlo software.

Expected Monetary Value

Earlier in this chapter it was mentioned that quantitative risk analysis further analyzes risks. Particularly, we spoke of the risks that exceeded our threshold in qualitative risk analysis. Table 5.4 shows those risks requiring more evaluation.

Risk #	Risk Title	Risk Description	Risk Ranking
2	Missile size	Can a missile be developed that is small enough to fit on a submarine and yet be able to fly 1500 miles as required? If not, this entire project may be at risk.	Red negative
4	Lack of a project management methodology	If a project management methodology cannot be developed to assist this project, it may not be possible to bring the project in on time or on budget.	Red negative
6	Expert knowledge	Having the best theoretical knowledge and experience in the Western world may allow us to minimize the need for contract consults thus reducing costs.	Red Positive
7	Government backing	The backing of the United States Government may allow us the resources to crash many more activities than previously thought, thus shortening the project.	Red Positive
8	Lack of project management skills	The team lacks project management skills, and they have not fully worked together. This may translate to poor management of the project.	Red negative

9	Lack of creativity	By being educated in advanced sciences, the project team may not have the creativity needed to bring the project in on time and on budget.	Red negative
11	Guidance system	A reasonable guidance system for the missile may not be easily and quickly developed, thus causing schedule delays.	Red negative
15	Presidential backing	If presidential backing can be maintained or increased, the possibility exists for increased funding to the project.	Red Positive

Table 5.4: Risks from Qualitative Analysis Needing Further Evaluation

One method commonly used to do this further evaluation is known as Expected Monetary Value or EMV. To use EMV, we again use an equation. This equation is shown below.

$$EMV = Impact \times Probability$$

This equation should look familiar from the last chapter. However, it is quite a bit different. In the last chapter, we used a similar equation to multiply two scales together such as 3 probability x 3 impact = risk rating or in this case 9. This number was used to prioritize the risks. For our use here, instead of using scaled numbers, we will be using actual values.

There are a couple of common methods used to obtain these actual values. One method is to determine the actual cost should a particular risk occur and determine as close as possible the actual probability of the risk occurring. This method may not be possible for all risks in a certain project. In fact, for some projects, it may not be possible for any risks.

If it is not possible to do this, we go back to our risk ratings from the last chapter, our risk matrix from risk planning, and our definitions of probability and impact. We know all risks that exceeded our threshold were red risks. Table 5.5 shows how each risk obtained its color ranking by listing its probability and impact.

Risk #	Risks	Proba-bility	Impact	Color Ranking
2	Missile size	Very High	High	Red
4	Lack of a project management methodology	Very High	High	Red
6	Expert knowledge	Very High	High	Red
7	Government backing	High	Very High	Red
8	Lack of project management skills	High	Very High	Red
9	Lack of creativity	Very High	Very High	Red
11	Guidance system	Very High	Very High	Red
15	Presidential backing	Very High	Very High	Red

Table 5.5: Risk Color Ranking Breakdown

From Table 5.5, we cross reference this to the definitions of impact and probability we created during risk management planning. Table 5.6 shows the "High" and "Very High" portions from that table.

Scale	Proba-bility	Schedule Impact	Cost Impact	Scope/ Performance/ Quality Impact
Very High	>80%	> 6 months	> 10 million	Catastrophic reduction in original Polaris/ SSBN program objectives. System unable to perform original function.
High	61-80%	2 months to 6 months	5 million to 10 million	Major reduction in original Polaris/ SSBN program objectives.

Table 5.6: High and Very High Portion of Definitions of Impact and Risk Table

To cross reference the two tables, we replace a high or very high in Table 5.5 with data from Table 5.6. Since the data on the table in some cases has a span between two values, one could take the average of the two to calculate EMV or one could be conservative and only use the highest value. For this example, I chose to use the conservative value to reflex the original risk averse stakeholder tolerance. For monetary values of Very High, a realistic impact value was used. These data could be obtained from historical records or other available information.

Risk #	Risks	Proba-bility	Impact	EMV
2	Missile size	.99	$10 million	$9.9 million
4	Lack of a project management methodology	.99	$10 million	$9.9 million
6	Expert knowledge	.99	$10 million	$9.9 million
7	Government backing	.80	$25 million	$20 million
8	Lack of project management skills	.80	$15 million	$12 million
9	Lack of creativity	.99	$20 million	$19.8 million
11	Guidance system	.99	$30 million	$29.7 million
15	Presidential backing	.99	$35 million	$34.65 million

Table 5.7: EMV Values

This table shows us that even though all the above risks were ranked as a red risks, clearly some risks are much, much more significant than others on this list. As one can see, EMV allowed us to further analyze and refine these risks and to create more appropriate responses.

Decision Trees

Decision trees are used to evaluate particular decisions made for a project. Decision trees employ Expected Monetary Value (EMV) to improve the value of the result. Decision trees calculate the EMV of different states of nature associated with the decision. In this way, much of the subjectivity is removed from the decision and replaced with objectivity.

Figure 5.9 is an example of a simulated decision tree from the *Polaris* Submarine Project. Early in the project, it was determined that if you plan on launching missiles from a submarine, you need a submarine. Not too surprising, is it? No such capable submarine existed in the world. That left two choices: build a new one from scratch or modify an existing submarine. After a quick search, it was noted that the U.S.S. *Scorpion* was being constructed in Groton, Connecticut. After some extensive research, it was determined that the hull of the U.S.S. *Scorpion* could be extended by 130 feet to accommodate missile tubes. The decision seems pretty straight forward, doesn't it? Use the *Scorpion*.

Here's where it gets a bit tricky. If we decided to build a new ship, the extra time required to design and manufacture the new ship should allow time to actually design a missile that would work with the new submarine. This would significantly increase the probability of a successful outcome to the project. However, if it proves impossible technologically to design such a missile, the new submarine would be virtually useless.

If, on the other hand, we use the *Scorpion* realizing that a missile has yet to be designed to be launched from such a ship, we might encounter extensive scope creep and massive rework to make the ship functional once a missile design is complete. This is especially true if we commence the modifications to the *Scorpion* before a missile or launch system is ready. Below is a decision tree to express this dilemma.

Figure 5.9 shows our decision on the left of our decision tree in the hexagon. Our two choices are shown as the two lines coming off the decision hexagon. Let's start with our first choice: modify an existing submarine. Modifying an existing submarine will likely cost at least $110 million dollars. This choice has three possible outcomes or states of nature. One is success. Success will cost the project no additional funds and has a probability of 60%. The second possible outcome is

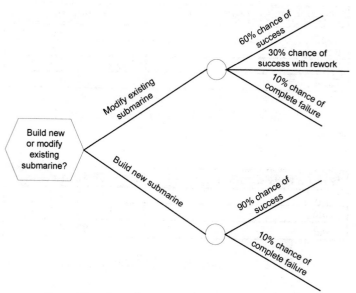

Figure 5.9: Sample Decision Tree for Polaris *Project*

considerable rework after modifications are begun. This state of nature will most likely cost $60 million dollars with 30% chance of occurring. The third state of nature is failure. Since no missile existed at this time that could perform as needed for success in this project, failure was always a possibility. The probability of failure is set at 10%. This would result in a loss of all funds spent in the original construction and in the modifications or $150 million dollars.

The second choice has two states of nature: success and failure of the new design. This choice will cost at least $220 million dollars to build a new submarine. Success as in the modification scenario will cost no additional money beyond the original $220 million. Failure, as seen above, is set at 10% and will also result in a virtually useless submarine. Now let's do some calculations.

Modify Existing Submarine	Outcomes
Success	.6 X 0 = 0
Rework	.3 X $60M = $18M
Failure	.1 x $150M = $15M
Total	**$33M**

Table 5.8: Sample EMV for Modification

Build a New Submarine	Outcomes
Success	.9 X 0 = 0
Failure	.1 X 220 = $22M
Total	**$22M**

Table 5.9: Sample EMV for New Submarine

From the tables above, we see the EMV of modifying an existing submarine would be $33M. Add this to the cost of $110M and we get $143M as the cost of this decision. From tables above, we see that the EMV of building a new submarine is $22M. Add this to the cost of construction and we get $242M. Clearly the choice should be to modify the U.S.S. *Scorpion* rather than build a new submarine for this project.

In reality, much of the decision to modify or to build a submarine was based on the Operational, Systems and Engineering Analysis performed by the China Lake Naval Weapons Station, California. They performed the detailed analysis to determine which decision was the more appropriate one.

With that said, the U.S.S. *Scorpion* (SSN 598) was renamed the U.S.S. *George Washington* (SSBN 598) and modifications began in Groton Connecticut. Until the U.S.S. *George Washington* was struck from the United States Navy's list of ships on April 30, 1986, its forward escape hatch had a plaque in it stating that the ship was the U.S.S. *Scorpion.* The *Polaris* Submarine Project now had a submarine.

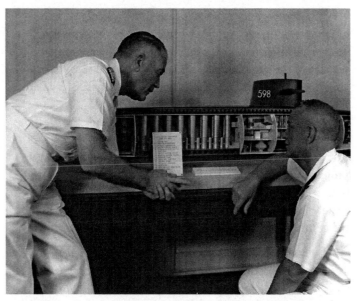

Figure 5.10: Admiral Raborn and Admiral Burke examining a model of the U.S.S. George Washington in July, 1959 showing 130 foot missile tube extension (Courtesy of U. S. Navy)

Contingency Reserves/Reserve Analysis

As mentioned earlier, contingency reserves are for those known-unknown risks. In many projects, there are two typical risk reserves: contingency reserve and managerial reserve. The managerial reserve is designed to provide a reserve for the project for what is known as unknown-unknown risks. These are risks that were not identified until they occurred. Hence, they were unknown as to existing and, hence, it is unknown if they might occur. The managerial reserve is normally computed as a percentage of the project budget. Common values for the managerial reserve are 5, 10, or 15%. Managerial reserves are normally controlled by senior management.

The contingency reserve is commonly calculated in one of two ways. It, too, can be a percentage of the project's budget. If this method is used, usually it is five or ten percent of the total budget. A second method to calculate contingency reserve is by the use of EMV. The EMVs of all risks that are not on the watch list are calculated. Negative values indicate negative risks; positive values indicate positive risks or opportunities. From this, we algebraically sum the EMVs. This value becomes the contingency reserve for the project. (See Table 5.10.) Stakeholder tolerance should also be considered when creating the contingency reserve. If the stakeholders have a lower tolerance of risk, one should increase the contingency reserve to compensate for this lower tolerance.

By examining the table, we see that we would want to have as a minimum $16.75 million in our contingency reserve. For a project of this magnitude and importance, it would probably be better to use the 10% of the total budget as the contingency reserve. But under no circumstances should we use a value less than $16.75 million. Depending on the company, contingency reserves can be held by the project manager, sponsor, or senior management.

Reserve analysis is a process that examines the risk associated with a given activity and provides appropriate schedule and cost reserves to compensate for this risk. In other words, if a given activity is more risky than another activity, reserve analysis provides that this activity should receive more budgetary and schedule buffers to accommodate this extra risk. This means the duration is usually modified with additional time and the budget is buffered with additional money to compensate for the additional risk.

Risk #	Risks	Proba-bility	Impact	EMV
2	Missile size	.99	$10 million	-$9.9 million
4	Lack of a project management methodology	.99	$10 million	-$9.9 million
6	Expert knowledge	.99	$10 million	$9.9 million
7	Government backing	.80	$25 million	$20 million
8	Lack of project management skills	.80	$15 million	-$12 million
9	Lack of creativity	.99	$20 million	-$19.8 million
11	Guidance system	.99	$30 million	-$29.7 million
15	Presidential backing	.99	$35 million	$34.65 million
			Total	**-$16.75 million**

Table 5.10: Contingency Reserve Calculation

Further, this additional reserve, as well as any contingency reserve and or managerial reserve, needs to be tracked throughout the life of the project. These are not bottomless pits that funds can be drawn from ceaselessly. A project manager must know at any time in project the remaining reserve funds for their project.

Figure 5.11: Polaris underwater test facility launch pad stool circa 1958. (Courtesy of G. Verver)

Figure 5.12: Polaris underwater test facility pad stool section in the shipyard, circa 1958. (Courtesy of G. Verver)

Figure 5.13: Polaris underwater test facility launch pad stool in the shipyard, circa 1958. (Courtesy of G. Verver)

Figure 5.14: Polaris test facility TV camera tower, Circa 1958. (Courtesy of G. Verver)

Figure 5.15: Polaris underwater pop up launch pad base in the shipyard 1958. (Courtesy of G. Verver)

Figure 5.16: Polaris underwater test facility TV camera tower being erected 1958. (Courtesy of G. Verver)

Figure 5.17: Polaris underwater test facility base and stool in the shipyard 1958. (Courtesy of G. Verver)

Figure 5.18: Polaris underwater test facility launch stool on the barge at San Clemente Island, CA in 1958. (Courtesy of G. Verver)

Figure 5.19: Polaris Missile Pop Up Tests at NOTS China Lake, San Clemente Island, CA in 1959. Much of the design, testing, analysis, and final system concept were done at China Lake Naval Base. (Courtesy of China Lake Museum)

Figure 5.20: March 9, 1960 Polaris Launch from Cape Canaveral, Florida (Courtesy of USAF)

## CHAPTER SIX

Risk Response Planning

Once risks have been analyzed by qualitative and quantitative methods, one must decide what to do about these individual risks. There are a number of choices available. For positive risks, strategies include exploit, enhance, share, or accept the risk. For negative risks, strategies include avoid, transfer, mitigate, or accept the risk. Employing these strategies, one creates risk response plans to address the risks we determined that we want to address.

Figure 6.1: Risk Response Planning as part of the overall process

There is no expectation that one creates a risk response plan for every risk. That would be too costly and not very productive. You want to limit yourself to risks that exceed the threshold you set in previous steps. Those risks with the highest EMV, risk ranking, or the ones in red on the matrix are usually selected to start with. You may not even have enough money to respond to all of these. Your risk budget plays an important part in your selection of which risks to address. Do not waste your time or budget on risks you previously placed on the watch list. You already decided not to address these risks any further, so spending your precious risk budget or time on these risks is counterproductive.

Utilizing the Pareto Principle, or 80/20 rule, implies that 20% of the risks will create 80% of the issues on a project. Concentrating on the top 20% of the risks, whether they be positive or negative, will provide the most impact for least cost. Attempting to create strategies other than acceptance for greater than 20% of the risks in your project is a losing proposition. The more money one spends on risks outside of the top 20%, the less and less value one gets back for each dollar spent. Thus, each dollar spent on sequent risks provides less value than the dollar before it did, and so on. Believe it or not, the Law of Diminishing Returns also applies to risk management.

Risk response planning does not end with simply developing plans that utilize these strategies. Risk owners also need to be assigned. Risk owners are those individuals or groups responsible for implementing these risk response plans. It is important to choose the right person or group as risk owners. You want to assign someone who can actually address the risk. This may be (and often is) someone not on the project team.

A risk response plan does not simply say "Enhance risk number 1.3.4." It must actually describe how one plans on implementing the chosen strategy. It must be in sufficient detail to allow the risk owner to successfully implement the plan. In many cases, one may have to implement more than one type of strategy or employ the same strategy more than once. These plans should also include when in the project the risk is likely to occur, costs if it should occur, cost of responses, assigned risk owners, and any other pertinent information.

Positive Risk Strategies

Exploit

Simply put, exploiting a risk means making the opportunity happen. It is such a good opportunity for bettering the project that you just have to make it a reality! Areas normally considered for exploitation are quality, schedule, and cost. If I can find a way to get the project done faster, or with better quality, or at a lower cost, I might consider exploiting the opportunity. Exploitation usually involves a significant cost. Getting the project done faster or with higher quality is not free. One needs to consider the cost of exploitation versus what may be the gain to the project before choosing this strategy. For this reason, exploitation is not often used as a strategy.

Enhance

When enhancing an opportunity, one attempts to increase its impact or its probability of happening. One cannot say for sure if the risk will occur, but overt efforts are made to help make it happen. This involves using the triggers identified for the risk and the risk's underlying causes to aid in improving its chances of occurring.

Exploiting or enhancing a risk should not be confused with gold plating. Gold plating is providing the customer more than they asked for as a way of enhancing customer satisfaction. Exploiting and enhancing are concerned with risk management, not with customer satisfaction. Gold plating is never an advisable technique for a project manager to use.

Share

Sharing a risk or opportunity means sharing the opportunity with another party. This is commonly done when one cannot make use of the opportunity with one's own current resources. In other words, I cannot do this on my own, but may be able to if I partner with someone else. I will then bring in another individual or company to share the opportunity and share in the benefits of the opportunity. This is usually done through a partnering agreement, a memorandum of understanding (MOU), or a contract.

Accept

Most risks on a project usually end up being accepted. Many are too low of a probability or too low of an impact to bother with, so they are simply accepted. This means you are willing to accept the consequences should the risk occur. Acceptance is a common strategy for risks on the watch list. Acceptance is used as a strategy for both positive and negative risks.

There are two types or methods of acceptance: passive and active. Passive acceptance means you have decided not to do anything overt about this risk. Basically the risk is placed on the watch list and revisited should it occur. In some cases, it is impossible to come up with any other acceptable strategy and one is forced to employ passive acceptance. In other cases, the risk is inconsequential, so why bother expending time and resources to worry about it.

The second type of acceptance is active acceptance. Active acceptance is different from passive acceptance in that contingency plans may be created in advance should the risk occur or contingency reserves are assigned to this risk in case it occurs. More on this topic can be found under Contingency Planning at the end of this chapter.

Negative Risk Strategies

Avoid

Avoiding a risk making sure that the risk will not occur. You eliminate the underlying cause or causes of the risk. This may involve changing the WBS, eliminating the work that would cause the risk, or modifying the scope of the project to ensure the risk will not occur. This may involve reworking the project objectives and requirements.

Avoidance should be reserved to those risks that, if they should occur, the consequences would be disastrous to the project. Avoiding a risk may require quite a bit of replanning of the project by the project team. It may also be somewhat costly. As with all risk management strategies, it is not usually cost effective to spend more to avoid a risk than the risk's EMV. Avoidance is not commonly employed on a project, as its cost in both time and money tend to be high.

Mitigate

With mitigation, you do not eliminate the possibility of the risk happening. Instead you reduce its probability, its impact should it occur, or both. Mitigation usually involves adding activities to the WBS such as prototyping, testing, moving the location of the work, or using vendors more qualified to do the work.

The risk may still occur when using mitigation. With some risks, it is not possible to eliminate the risk without losing an important part of the project or without spending a significant amount of money. It is usually better to employ a mitigation strategy on these risks. When considering mitigation, you must first assess if it is possible to reduce the risks probability, impact or both. If it is possible to address one of these, then mitigation is probability a good choice. Again one should limit the amount spent on mitigation strategies to the EMV of the risk.

Transfer

Transference involves transferring the risk to a third party. As you are not eliminating the risk, the risk may still occur, so you may still have consequences to deal with. You are not necessarily reducing its impact or probability. What you are doing is transferring the management of the risk to someone else.

Transferring a risk can be done using various methods. One of the most common forms of transference is hiring contractors or outsourcing. Through the contract, the risk is transferred to the contractor. The contractor now becomes responsible of the management of the risk and to some degree its consequences. There are numerous types of contracts available. Each type of contract assigns a different amount of risk to the contractor and to the buyer of the services. Be sure you use an appropriate type of contract for the work you want done. Remember, contractors will want to be well compensated for the risks they are assuming, so contracting may be expensive. Contracting should be used if you do not have the expertise to do the work, the technical ability to do the work, the resources to do the work, or you do not have the free capacity to do the work. If trade secrets are involved in the work you are considering outsourcing, you might want to think twice about contracting or outsourcing it, as you may not want to share these secrets with others. When considering contracting or outsourcing, ask yourself: will it cost more for me to do the work or will it cost more to hire a contractor to do the work?

Another common method of transference is insurance. You may be able to buy insurance for various risks associated with your project. Insurance typically protects you financially from the risk. The risk may still occur, but if it does, you are at least partially protected by the insurance from financial loss. Performance and surety bonds are common forms of insurance on projects. Warranties and guarantees are also considered forms of risk transference.

Accept

Acceptance of a negative risk is performed in the same manner as that of a positive risk. Since this acceptance was explained in the positive risk section, it will not be repeated here.

Contingency Planning

Contingency planning involves creating plans to follow should certain risks occur. It does not eliminate the risk, nor does it reduce its impact or probability. These plans are put in action after the risk has already occurred, thus they are reactive, not proactive. These plans are created during the planning phase for use later during the execution of the project. In that way, once these risks occur the project manager does not need to scramble around trying to develop a response. The project manager simply dusts off the preconceived contingency plan and implements it.

These plans are normally paid for out of the contingency reserve mentioned earlier. These plans are created to offset threats posed by these risks on the project schedule, budget, scope, or quality, should these risks occur. Contingency planning can be used for risks for which other strategies are not appropriate solutions or where other strategies cannot lower the potential threat to an acceptable level. Not all risks will have contingency plans. As we saw earlier, risks with a low probability and low impact may not have any strategy at all and are simply accepted.

Residual and Secondary risks

One important point to remember when performing risk response planning is that it is highly likely you created new risks as a direct result of the responses you implemented to offset existing risks. In other words, by creating and implementing a risk response, you likely created new risks. Your risk response may open a whole new can of worms where risks are concerned. These new risks are called secondary risks. One needs to review response plans from the viewpoint of "Did I create any new risks by this risk response?" You need to redo some risk identification after creating your risk responses. Once identified, these new risks also need to be analyzed just as all other identified risks were analyzed earlier. And, yes, this may require creating still more risk response plans.

A second point to remember is the risks that still remain after all plans are created are known as residual risks. As stated earlier there is no expectation that any project manager will eliminate all risk on a project. The only way to do that would be not to do the project at all. This is usually not a good idea. Furthermore, all of these strategies, with the exception of passive acceptance, incur a cost. If you try to do too much in the way of risk responses, your budget for risk management may soon exceed your budget for the project – not a good position to be in. Risk response is always a question of how much can I do to minimize risk with the money I have available. Even if I choose to mitigate versus avoid risks in order to save money, residual risks always remain when employing mitigation as a strategy. Remember: mitigation does not prevent the risk from occurring, thus residual risk remains. All this being said, it should be clear that risks will remain, after all is said in done. Residual risks are not an unusual entity in project risk management.

Contracts

We discussed contracts in detail when we examined transference as a strategy. Risk Response also provides one more important facet to contracts: contract language and type. In addition to deciding to transfer, share, or mitigate a risk, one need consider what should be included in the subsequent contract. Once risks have been analyzed and responses created, one should consider what wording should be added, modified, or deleted from contracts in response to these risks. Based on risk analysis and responses, new contract terms and conditions may need to be created to reflect the current state of risk on a project. A new contract statement of work may need to be written or a proposed one may need to be modified based on new knowledge concerning risks. The choice of contract type to be used should be based on the perceived risk on a project.

Three types of contracts are commonly used in project management: fixed price, time and materials, and cost reimbursable. Cost reimbursable contracts are also known as cost plus contracts. Figure 6.2 shows the various contract types versus the risk associated with each. In Figure 6.2, the initials CR represent a Cost Reimbursable contract, T&M represents a Time and Materials Contract, and FP represents a Fixed Price Contract.

As shown in the figure, a fixed price contract transfers much of the risk from the buyer on to the seller's shoulders. In this type of contract, the contract risk is the least for the buyer and most for the seller. In a cost plus contract, the risk is firmly placed on the buyer's shoulders with minimal risk to the seller. A time and materials contract provides a somewhat equal sharing of risk between the buyer and the seller. One word of warning: it is not advisable to use a fixed price contract on advanced technical programs, as there are simply too many variables that may impede success.

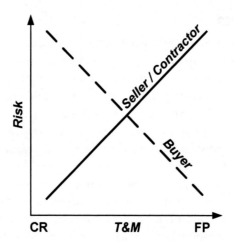

Figure 6.2: Simple Contract Types Versus Risk

These general types of contracts commonly have subtypes. For instance, in order to synchronize the priorities of the seller with those of the buyer, incentives and award fees may be added to fixed price and cost reimbursable contracts. If we are discussing a fixed price contract, adding incentives creates a FPIF – or fixed price with incentive fees. Incentives provided monetary benefits for the seller to achieve a particular goal of the buyer such as completing the project by a certain date or by achieving a certain quality level. Awards fees provide monetary incentives to achieve a particular goal within a project such as "complete phase one by July first."

Here are some important good practices when working with contracts. First, have only authorized individuals negotiate and administer the contract. You want to minimize the number of individuals that "speak for the company." This will reduce confusion in communications between the buyer and the contractor. Second, it is critical that all correspondence concerning the contract should be in writing. This includes change requests, approved changes, contract modifications, and any nonconformance with the terms of the contract. It is important to maintain a "paper trail" in case one is needed in the future. Third, evaluate all change requests for schedule, cost, and scope impact before approving the change. Don't just approve a change request to the contract because the contractor requests one. Lastly, know the contract. It is a sure bet the other party does!

Polaris Program Risk Strategies

Now that we covered the options for positive and negative risks, let's come up with response plans for our red ranked risks. Below is our risk register from our quantitative risk analysis.

Risk #	Risks	Probability	Impact	EMV
2	Missile size	.99	$10 million	-$9.9 million
4	Lack of a project management methodology	.99	$10 million	-$9.9 million
6	Expert knowledge	.99	$10 million	$9.9 million
7	Government backing	.80	$25 million	$20 million
8	Lack of project management skills	.80	$15 million	-$12 million
9	Lack of creativity	.99	$20 million	-$19.8 million
11	Guidance system	.99	$30 million	-$29.7 million
15	Presidential backing	.99	$35 million	$34.65 million
			Total	**-$16.75 million**

Table 6.1: Risks Requiring Response Strategies

Risk number two: Missile Size tops our list. The requirement of the *Polaris* Submarine Project was a submerged launch ballistic missile with a range of at least 1500 miles. The risk is that such a missile may not be possible. This is definitely a negative risk. It would be impossible to avoid this risk. We cannot simply make it go away. Mitigation would be equally difficult. How can I lessen the probability or impact of such a risk? Transference is also out of the question. We are the experts. If we cannot figure it out, who could we possibly transfer the risk to? We already know we will not be accepting this risk. It is simply too dangerous to the project to accept.

That leaves the creation of a contingency plan. And this is exactly what happened. A plan was created as a fall back plan. If it was not possible to develop a 1500 mile missile, then could we create one that flew 1000 miles? There had to be a minimum range for the missile for it to be useful. The contingency plan created a point in time at which we would determine if such a missile was possible or not. If such a missile was not possible, then we would begin designing a shorter range missile that was possible. Once this missile was complete, we would then go back and continue work on the original 1500 mile design. In that way, we could complete the project on time and upgrade the missiles later in another project. Such a point was reached in the late 1950s, and the missile range was reduced. In fact, it actually took until 1963 to develop a missile with a range of 1500 miles. This was due to the propellant technology available at the time. But for now, *Polaris* could continue, albeit with a shorter range missile.

Risk number four, lack of project management methodology, is second on our list of those risks requiring response strategies. It was critical that the *Polaris* Submarine Project be completed on time. It did not take long to realize that without some methodology, this would be impossible. Developing such a methodology by the project team was not very likely, as many on the team were, well, rocket scientists (and rocket engineers).

Going though our choices of possible strategies, we can exclude acceptance out of hand. We would never get the project done on time if we simply accepted this risk and did nothing. Avoidance would be equally difficult. Since no methodology currently existed, how could one come up with an avoidance strategy? Mitigation is not a good choice either, as how can non-experts in the field reduce the probability or impact of something they are not experts at?

That left transference as the team's only remaining option. And that is the solution they used. The team decided to obtain the assistance of technical experts from the Project Evaluation Branch - Special Project Office of the U.S. Navy. The Special Projects Office then brought in Lockheed Missile Systems and Booz-Allen & Hamilton to create PERT. In this way, the project transferred some of the responsibility for this risk to outside organizations. This project might get done on time after all.

Let's examine risks eight and nine at the same time. Risk number eight is the lack of project management skills. Risk number nine is the lack of creativity and possible technical disputes within the project team. Both of these are negative risks. Both affect the ability of the team to efficiently manage the project. Certainly acceptance is out of the question. The project cannot be successful if we do not address these two risks. Transference is a possible answer, as it may be possible to hire project management consultants to manage this project. There are a number of issues with this approach. For one, there is a question of security. The more people with knowledge of the project and who work on the project, the more difficult it is to maintain security on the project. Also, do you really want to bring in someone from the outside of the project to manage a project of this importance? Probably not! Avoidance is also not possible. There is nothing one can do to totally make these risks go away other than, of course, canceling the project.

That leaves mitigation as a risk response strategy and that is what happened. These risks were quickly recognized by the project's sponsor: Admiral Arleigh A. Burke. Admiral Burke at that time was the United States Navy's Chief of Naval Operations (CNO). To mitigate these risks, Admiral Burke took two major steps. First, he created the Special Projects Office (SPO). This office was designed to manage and support this project. It was to be a new type of organization, as opposed to the traditional Naval organizations consistently used in the Navy. The use of this new organization allowed for more creativity and better project management, since it has a less complex structure. Second, Admiral Burke appointed Admiral Raborn as the Director of the Special Project Office, making him the project manager of the project. Admiral Raborn was the type of leader that was not necessarily a technical expert in any area of the project, but he was known for getting things done. Actually, Admiral Raborn was a carrier man. He was awarded the Silver Star for his actions while stationed as the executive officer on the U.S.S. *Hancock* after it was damaged by a Kamikaze in World War II. Admiral Raborn was a determined, driven leader. If someone could do this project on time, it was Admiral Raborn.

As with the last two risks, let's examine numbers seven and fifteen together. Risk number seven is governmental support, and risk number fifteen is presidential support. Both of these risks are positive risks. Both of these are critical to the success of the project. Hence, acceptance is not advisable. Exploitation is also not possible, as it is not possible to do anything to absolutely guarantee presidential or governmental support of anything. That leaves only sharing and enhancement. Who do you share this opportunity with, the Soviets?

This leaves enhancement as the remaining choice. But how do you do enhancement? This one is very difficult. As a naval officer, one cannot dictate to government or the president. Admiral Burke understood this. Instead, he enhanced these risks in a number of ways. First, he reported to congress and the president on the successes of the project. Positive press is never a bad thing. Second, he transferred funding away from other projects to this one. In other words, he ran interference for this project. Before Admiral Burke transferred these funds, the *Polaris* Project was just one of many projects the Department of Defense was doing. It was underfunded and had a low priority.

Initially, the U. S. Navy was not thrilled to be part of this project. It did not see its role as being relegated to that of a bunch of ships hiding from the world toting around ballistic missiles. It was currently developing *Jupiter* cruise missiles for at sea launch. It did not want to spend what little funds it had on a new ballistic missile platform.

It was also falling victim to interservice rivalry. The Navy had recently lost out to the U.S. Air Force in the B-36 versus aircraft carrier controversy. This turned out to be major struggle over military funding priorities. By losing to the Air Force, the Navy had to cancel building an aircraft carrier, so the Air Force could build more B-36 Peacemaker Bombers. This led to the Admirals' Revolt of 1949 in which the Navy's admirals feared for the very survival of the navy. It did not want to enter into another conflict with the Air Force – this time over ballistic missiles. Initially, this project had little chance of success as it had little governmental support. Admiral Burke's actions may not have increased government backing or presidential backing, but it did, at least, ensure funding for the project.

Some of the eventual governmental and presidential support for this project did not arise from anything done by the *Polaris* Submarine Project. In 1959, the Soviet Union's SS-N-4 Ballistic Missile began its flight test program prior to it obtaining operational status – thus making it the first submarine ballistic missile system. This was not a submerged launch system; to fire this missile, the submarine had to surface. The missile had a range of 350 miles and carried a nuclear warhead. This external event, combined with the launch of Sputnik, did much to further raise the priority of *Polaris*.

A second boost to presidential and governmental support was the election of President Kennedy. President Kennedy campaigned on the "missile gap" with the Soviet Union. This was a gap between the United States and the Soviet Union in which the Soviets were perceived to have had more and better missiles than the United States possessed. In fact, no such gap existed. He also campaigned on the low defense budgets of President Eisenhower. These two points may have contributed to the election of President Kennedy, but these two historical events certainly contributed to the success of *Polaris*.

On to the next risk: number six – Expert Knowledge. At the time, we had some of the greatest scientists, physicists, engineers, and technicians in the world working on this project. How can we utilize this risk to the benefit of the project? We could accept this risk and hope for the best. Since it is a positive risk, it would not hurt the project to accept this risk, but it would not help the project either. How would one share this risk? Who would you share it with? Exploitation is not a good choice either. I would love to find a method to positively guarantee the proper use of the expert knowledge on this project or any project; I am not sure it is possible.

Enhancement is a possibility. We could enhance the probability or impact of this risk occurring. That would be a better solution than doing nothing and hoping for the best. So let's exclude acceptance and focus in on enhancement. As shown above, this is what Admiral Burke did. This was one of the purposes of the Special Projects Office. By creating a vertical organization, one could enhance, leverage, and better utilize the expert knowledge of the project team.

Our last remaining risk is risk number 11: the guidance system. The guidance system meant everything to the project. We could build a submarine, we might be able to build a nuclear ballistic missile, and we probably could marry the two by building a launch system. But none of this matters if we cannot guide the missile to a target! We are obviously not going to have someone pilot the missile. This excludes acceptance of this risk. Equally, we cannot avoid the risk. We need a guidance system. So that leaves us with transference and mitigation. In this case, they used both. The project needed the best brains available to solve this problem. Contractors were brought in to work with existing Navy laboratories to design and test internal navigational guidance systems. Next let's talk mitigation. In order to mitigate this risk, Admiral Burke gave Admiral Raborn an out. If at any time it became apparent the project could not meet its objectives, Admiral Raborn was empowered to recommend to Admiral Burke to kill or modify the project. This minimized the overall risk to the United States Navy should the project prove impossible. In this way, the impact of this risk was mitigated.

When the Special Project Office was created by Admiral Burke, as mentioned earlier, the Army and Navy were jointly developing the *Jupiter* Missile to satisfy the requirements of this project. This involved modifying the *Jupiter* Missile for seaborne use. Many wondered if this was even possible. Once Admiral Raborn arrived on the scene as the Director of the Special Projects Office, it was quickly realized the *Jupiter* Missile would not work in a naval environment. For a start, it was liquid fueled. This highly volatile fuel, liquid oxygen, was extremely dangerous to have on board a warship. Second, this missile was physically too large and too heavy to be carried aboard a ship. Because of this, the Navy withdrew from the *Jupiter* Project. Further, solid fuel technology was not advanced enough to propel a missile of the size needed to meet the program's requirements. This left the Navy, technically, without a missile. We now have yet another risk. To mitigate this risk, Admiral Raborn appointed Captain Levering Smith to the position of Technical Director, Special Project Office, *Polaris*. Captain Smith was one of the Navy's top experts in solid fuel propulsion systems. He could understand and could solve technical problems. Captain Smith was given the responsibility and resources to solve the missile fuel dilemma. Captain Levering Smith made if happen! *Polaris* now had a technical leader.

As just noted, Captain Smith was an expert in solid fuel rockets. So, how does he solve such a problem? First, Captain Smith engaged the China Lake Naval Weapons Station to study what size a liquid fueled rocket would have to be to have the range required by the project. This was done first, since much more was known about liquid fueled rockets. It was determined that a liquid fueled rocket would have to weigh five times that which was feasible and be twice as large as a submarine could carry. Second, Lockheed and Aerojet Corporation were engaged to study what a solid fuel rocket could do. At about the same time, China Lake Naval Weapons Station evaluated what size warhead was really required to provide reasonable nuclear deterrence. In other words, how big did a nuclear warhead need to be to be effective? Up to this point in time, when it came to nuclear warheads, the philosophy was the bigger the better, but was this actually true? From all of these studies, it was determined that if a warhead could be made small enough to do the job, all of this was possible. Smaller, more accurate warheads were actually better than larger, less accurate warheads; they produced less collateral damage. Hence, the size of the warhead became a risk to the project.

These were not the only "new" risks to develop as time went on. Some other risks included: integrating the warhead on to the missile and reentry of the warhead into the atmosphere. Was technology advanced enough to allow the marrying of a nuclear warhead to a missile? It had never been tried – was it even possible? Second, once the nuclear warheads are in outer space, how can they reenter the atmosphere without being destroyed by the atmosphere? A heat shield had to be developed to ensure the safety of the warhead. All these risks were critical to the success of the project. To evaluate these risks, China Lake Naval Station was brought in to do a risk analysis for the entire project. This began as a thorough operations and systems analysis with the help of the Lawrence Livermore National Laboratory to determine what was possible given the projected technology and the projected deployment date of the submarine. From this, better requirements and a better vision of what *Polaris* was capable of were created. Captain Smith also inherited the responsibility to solve these risks. Admiral Rabon was not a technical expert in this area, but Captain Smith was. The combination of the two was critical to *Polaris'* success. Table 6.2 shows the new risks added to the risk register.

Risk #	Risk Title	Risk Description
32	Nuclear warhead size	Could a nuclear warhead be developed that was small enough to allow the use of a small solid fuel rocket? If not, this project may be in jeopardy.
33	Warhead integration	Is it possible to attach a nuclear warhead onto a missile? If not, what type warhead was possible? This risk poses grave schedule and cost issues for this project.

34	Warhead reentry system	Can a heat shield be developed to safely protect these warheads? If not, this risk poses significant schedule risk to the project.

Table 6.2: New risks added to the Risk Register

Earlier in this text, secondary risks were mentioned. Secondary risks were those risks created by the implementation of risk response plans for other risks. One response plan created above could have created a number of significant secondary risks. That response plan was the engaging of contractors to assist in the development of the internal guidance system. With each additional person brought in to the project, the security risks to the project increased. The United States was not the only nation working diligently to develop an internal guidance system. A few nations may have been willing to pay large amounts to money to obtain such technology. This posed a significant security risk. Another secondary risk that arose from engaging contractors to support the project is how the in-house Naval Laboratory employees would work with highly paid contractors? Would there be friction between the two groups, thus slowing the project down? Both of these two secondary risks would have to be evaluated to determine if response plans were warranted.

With response strategies in place for our significant risks, it was now time to move on to risk monitoring and control. Table 6.3 shows the secondary risks added to the risk register.

Risk #	Risk Title	Risk Description
35	Project security	With every additional individual added to the project, maintaining project security becomes more difficult. If project security cannot be maintained, all that is created may be lost to countries outside of the United States
36	Friction between in-house and contractor workforces.	As more outside contractors are added to the project, the probability of friction between in-house and contract personnel increases. Such friction could pose significant schedule risk to the project.

Table 6.3: Secondary Risks added to the Risk Register

## CHAPTER SEVEN

Risk Monitoring & Control

Previously, we identified risks, we analyzed those risks through qualitative and quantitative methods, and we decided on an appropriate response for each risk. That ended the planning for risks. Now we move on to the next phase of our project. Risk monitoring and control occurs alongside project execution – when the work of the project is being preformed. It does little good to spend all this time and effort on risk planning if we do not monitor and control for risks during project execution.

Figure 7.1: Risk Monitoring & Control as part of the overall process

Risk monitoring and control is a constant process throughout project execution. Risk monitoring and control should be a major component of project status meetings. Risk monitoring and control should contain the following:

- Verifying that contingency plans created earlier are put in effect when needed and are effectively addressing the associated risk.

- Monitoring risk triggers to see if any risks are about to happen or are just starting to happen. If so, are contingency plans available to compensate for these risks?

- Updating the contingency reserve budget to ensure the contingency budget reflects the actual contingency reserves remaining.

- Ensuring the risk management process is being followed.

- Evaluating the effectiveness of the overall risk management process.

- Verifying that risk strategies are effective at controlling risks. If not, do conditions warrant creating new plans?

- Monitoring project performance measurements to see if new, previously unknown risks might be occurring. What is the likelihood that you identified each and every possible risk the first time through? Is it possible you might have missed a few? Earned Value Measurement, variance measurements, as well as quality measurements are good at finding new risks. Missing milestones and Earned Value indexes less than .9 or greater than 1.2 may be indicative of risks occurring. Monitor these measurements closely as a first line of defense against previously identified and new risks alike. When you notice one of these measurements going off course and you do not have any good answer for why, it could be an unidentified risk occurring. You might want to do some root cause analysis to identify the risk causing the off course measurements. Any new risks identified need to be evaluated, as all other risks were, to see if their impact warrants a response. Responses for new negative risks are commonly called workarounds. Creation of a workaround may require performing qualitative and quantitative risk analysis on a newly discovered risk. This is followed by risk response planning if required. For more on Earned Value Measurement, see Appendix A.

- Reviewing risks on the watch list to see if any are likely to occur soon or are occurring. One should also evaluate the actual impact of these risks should they occur in case a response is warranted.

- Monitoring secondary and residual risks identified earlier to evaluate if new response plans are needed.

- Updating the status of risks. This includes closing risks, deleting risks, and creating new response plans should conditions warrant.

- Performing risk reassessments on a routine basis. Review the impact and probability of risks to see if these are still valid. Have conditions in the project changed such that the original impact and probability ratings are invalid? Has the impact of the risk significantly changed either up or down? Are the risk responses created earlier still valid? Are the assumptions we associated with these risk still valid? As one can see, a new response plan may be needed for given risks. Risk reassessments also include reviewing the risk prioritization created during qualitative risk analysis. Due to current conditions in the project, should a given risk move up or down on the list, thus creating a heightened sense of urgency concerning the risk, or creating a reduced sense of concern should the risk move down on the list? A risk reassessment should be a regularly scheduled event. This ensures that the reassessment process occurs and is assigned appropriate importance in the project.

- Ensuring risk audits are being performed. Generally, it is a good idea to have an independent third party audit the risk management process. This third party should be specifically trained in performing risk audits. With all that is involved with risk management, it is conceivable that something might be missed by the project team. Auditors should audit the risk management process – especially the response plans and their effectiveness. In this way, they assist the project team in risk management.

- Lastly, ensuring the risk register is updated with any changes made during the previous steps.

For the *Polaris* Submarine Project, overall monitoring and control of the project was performed by the Project Evaluation Branch of the Special Project Office of the U.S. Navy. This Project Evaluation office, Admiral Raborn, Captain Levering Smith, and Admiral Burke worked hand-in-hand to make this project a success. From this project, the Special Project Office gained an international reputation for the effectiveness and innovation of its project control system. The results of their work are shown in the next chapter.

## CHAPTER EIGHT

Conclusion

Thanks to the work of thousands of dedicated individuals and the use of project management techniques such as PERT and project risk management, the Polaris Missile Program and the SSBN Submarine Programs were both highly successful. These two programs have been compared to the Apollo Program in scientific, engineering and technological achievements. As with the Apollo Program, these programs had to overcome seemingly impossible scientific, engineering, and technological obstacles, such as those described in this book.

"The early development of the U.S. Navy's Fleet Ballistic Missile, Polaris, was one of the most efficient Research & Development, Testing and Evaluation programs (RDT&E) and operationally effective strategic weapons systems ever accomplished by the DOD."
– Frank Knemeyer, Deputy Technical Director, US Naval Weapons Station China Lake, CA (Retired).

One of the more stunning achievements of this program was the development of the Ships Inertial Navigation System (SINS). This system allowed a submerged submarine to precisely navigate while staying submerged for weeks or months at a time. One could not accurately fire a Polaris Missile unless one knew where one was at when firing – unless you want to fire blindly in the dark. This system was pivotal for the program. Another remarkable scientific achievement was the development of a life support system for the submarine crew. This life support system generated the oxygen the crew breathed and distilled the water the crew drank. Without this achievement, the Polaris Submarine would not have been able to remain undetectable and submerged for months at a time. This allowed the food supply to be the major limiting factor in the longevity of individual missions.

Further, a new, compact warhead designated W-47 was developed by the Lawrence Livermore Laboratory just in the nick of time to allow the project to continue. Also, the China Lake Naval Weapons Station designed a system for integrating this new warhead into the missile and a heat shield for the warhead to allow the warhead to successfully make reentry into the earth's atmosphere.

Much credit for the success of the Polaris Submarine Program can be attributed to Rear Admiral W. F. Raborn, Captain Levering Smith, and Admiral Arleigh Burke. When Admiral Raborn was appointed head of the Fleet Ballistic Missile Program, later known as the Polaris Submarine Program, it was a program without a working missile or a submarine to launch the missile from, and not much in the way of funding to fix either issue. Sputnik – and help from Admiral Burke – changed all that. Admiral Raborn and Admiral Burke had the vision to realize that without project management, their task was hopeless. Admiral Raborn had the drive to steward the project's development. Captain Smith completed the leadership triad with his expert knowledge of the technology involved. Through it all, Admiral Raborn and Captain Smith managed to complete the program three years ahead of its original schedule. Quite an accomplishment indeed!

Later in his career, Captain Smith rose to the rank of Vice Admiral and led the Special Project Office. After Polaris, Admiral Raborn became the seventh Director of the CIA. Lastly, Admiral Burke retired as the U.S. Navy's Chief of Naval Operations on August 1, 1961.

Figure 8.1: Rear Admiral Arleigh A. Burke, Future Chief of Naval Operation, U.S. Navy (Courtesy U.S. Navy)

Thanks to his work and that of countless others, the U.S.S. George Washington was arguably considered the single submarine that most influenced world events in the mid to late 20th century. The mere threat of an undetectable weapon that could launch nuclear tipped missiles at an enemy thousands of miles away was enough to influence world events. This threat still influences world events today.

Figure 8.2: Rear Admiral Levering Smith – Technical Director, SPO. As a Captain, he led the design, development, and deployment of the Polaris A1 Missile (Courtesy of U.S. Navy)

Figure 8.3: Vice Admiral W.F. Raborn, Director, Special Project Office (Courtesy U.S. Navy)

On June 9, 1959 the U. S. S. George Washington SSBN-598 was launched from the builder, the Electric Boat Division of General Dynamics, in Groton Connecticut. She was commissioned December 30, 1959. About six months later on June 28, 1960, she sailed from Groton, Connecticut for Cape Canaveral, Florida. On board, she was loaded with two Polaris missiles. On July 20, 1960 she successfully launched both Polaris Missiles while submerged in the Atlantic Ocean off the coast of Florida. The launches were approximately two hours apart. The second missile traveled 1100 miles to an impact area guided by its own internal navigation guidance system. Figure 27 shows the message sent by the captain of the U.S.S. George Washington SSBN-598 to President Eisenhower of the successful launch of the Polaris Missile. History had been made.

Shortly thereafter, she departed for her first strategic missile patrol on the fifteenth of November that same year. Before she returned from that patrol, a second missile submarine was already at sea, the U.S.S. Patrick Henry (SSBN 599). Ballistic missile submarines from many navies still patrol the seas today.

Figure 8.4: U.S.S. George Washington (SSBN 598) Loading Polaris Missiles in Charleston, South Carolina (Courtesy U.S. Navy)

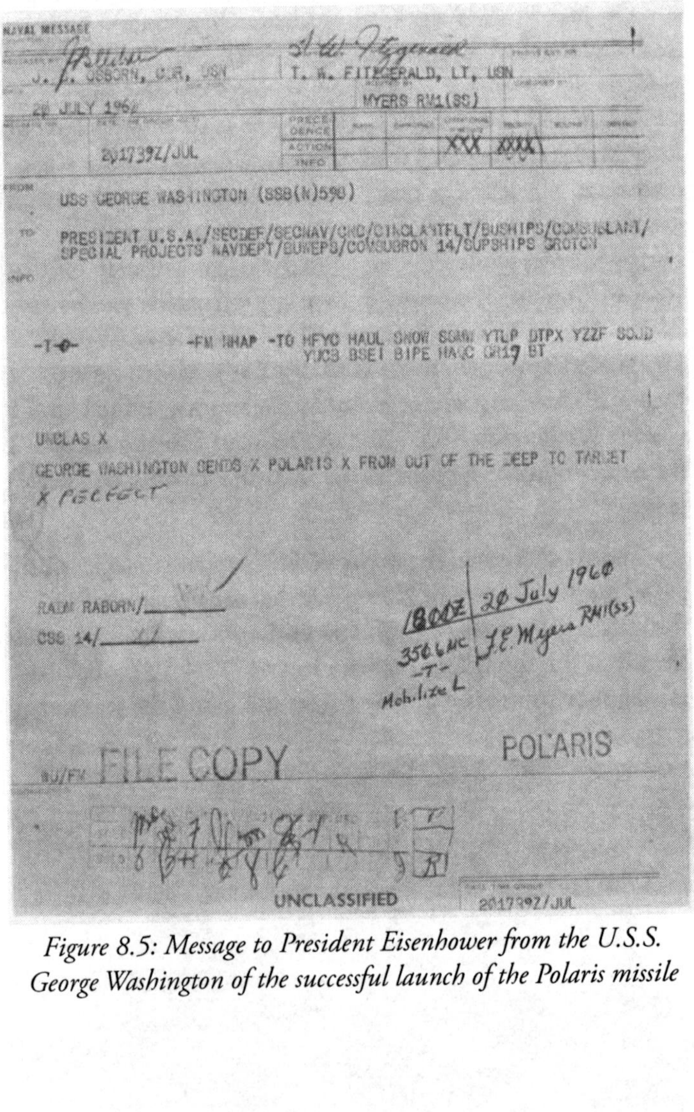
NAVAL MESSAGE

OSBORN, CDR, USN — T. W. FITZGERALD, LT, USN

MYERS RM1(SS)

20 JULY 1960

201739Z/JUL

XXX XXXX

FROM USS GEORGE WASHINGTON (SSB(N)598)

TO PRESIDENT U.S.A./SECDEF/SECNAV/CNO/CINCLANTFLT/BUSHIPS/COMSUBLANT/ SPECIAL PROJECTS NAVDEPT/BUWEPS/COMSUBRON 14/SUPSHIPS GROTON

INFO

-T-Ø- -FM NHAP -TO HFYC HAUL GNOW SGMN YTLP DTPX YZZF SOJD YJC3 BSET BIPE HAUC GR19 BT

UNCLAS X

GEORGE WASHINGTON SENDS X POLARIS X FROM OUT OF THE DEEP TO TARGET X PERFECT

RADM RABORN/

CSS 14/

1800Z 20 July 1960

356 LMC J.E. Myers RM1(SS)

-T-

Mobilize L

NU/FX FILE COPY

POLARIS

UNCLASSIFIED

201739Z/JUL

Figure 8.5: Message to President Eisenhower from the U.S.S. George Washington of the successful launch of the Polaris missile

Figure 8.6: Admiral Arleigh Burke aboard the U.S.S. Patrick Henry (SSBN 599) off Cape Canaveral, Florida for a Polaris Missile launch September 22, 1960 (Courtesy U.S. Navy)

Figure 8.7: A modern day version of a Polaris Submarine, the U.S.S. Rhode Island (SSBN 740) leaving on patrol. (Courtesy of the U.S. Navy)

Earned Value Management

E arned Value Management is a method of monitoring and controlling a project based on the amount of work completed, the actual costs, and the planned costs. Using these three values, one can determine whether a project is behind schedule, on schedule, or ahead of schedule. It can also determine if a project is on budget, over budget, or under budget.

To use this methodology, three values are calculated for each process/activity on a project. These values are the AC or actual costs; PV or planned value; and EV or earned value. The AC represents cost to date for an activity. This information comes from time sheets, invoices, material purchase orders, or other accounting data. The PV is how much was planned to have been spent by this time in the project on this particular process or activity. From these values and the formulas in Table A.3, we can calculate Cost Variance or CV; Schedule Variance or SV; Cost Performance Index or CPI; and Schedule Performance Index or SPI. When combined with all other activities, one can calculate the project's Estimate to

Completion or ETC; Estimate at Completion or EAC; and Variance at Completion or VAC. From these values, we get a complete picture of the status of the project.

Let's do an example. Let's begin by evaluating an activity. Assume the following information for activity A:

Activity A:

Cost to date = $500,000

The activity is 1/3 done

We are at the beginning of week 2 of 3

The original budget was set to be $1.5 million

From this data, we see the AC = $500,000, as this is the given cost to date. The EV is calculated by multiplying the percent complete by the original Budgeted cost. This is also known as BAC for this activity. Hence, .33 times $1.5 million is $500,000. The PV is calculated by how much should have been spent by now. Thus, at the beginning of week 2 of 3 means we have completed only one week out of three. This means we are 1/3 of the way through our planned scheduled time. We multiply this by our BAC for the activity and get $500,000. Using the formulas from our formula table for CV, SV, CPI, and SPI, we can determine the status of the activity. In this case both SV and CV are zero, indicating that the project is on budget and on schedule. Further, both the CPI and SPI are one also indicating that the project is on time and on budget.

To further illustrate Earned Value Management, let's add three more activities to the project. Below are the data for these other activities.

Activity B:

Cost to date = $1.6 million

The activity is 65% done

We are at the beginning of week 3 of 5

The original budget was set to be $2.5 million

Activity C:

Cost to date = $750,000

The activity is complete

The activity took 5 weeks to complete

The original budget was set to be $800,000

Activity D:

Cost to date = $1.5 million

The activity is 25% done

We are at the end of week 2 of 7

The original budget was set to be $3.5 million

Again we start by determining AC, CV, and PV. For activity B, we see the AC is $1.6 million. The EV is .65 times $2.5 million. This calculates out to $1.625 million. The PV is calculated by multiplying 2/5 by $2.5 million. The 2/5 is derived for having complete two weeks of the five weeks planned. This works out to $ 1 million.

Moving on to activity C, we see this activity is finished. Since the activity is finished, the PV and EV are the same value. In this case, $800,000, as this is the BAC for the activity. For all completed activities the EV=PV=BAC for the activity. The AC for this activity is given as $750,000.

Lastly, we calculate the values for activity D. The given AC is $1.5 million. The EV is 25% times $3.5 million or $0.875 million. The PV is 2/7 times $3.5 million or $1 million. Note the PV uses 2/7 since we are at the end of the week.

To calculate the Earned Values for the entire project, we next need to add the PV values together, the EV values together, and the AC values together. We also have to calculate the BAC for the entire project by adding all the individual planned budgets together. The values are show below as ACc, PVc, and EVc indicating cumulative values. The results are shown below.

Term	Value
ACc	$4.35 million
EVc	$3.8 million
PVc	$3.3 million
BAC	$7.3 million

Table A.1

Plugging these values into the equations below we get a more complete status of the project. These calculated values are show below. Note these values are for the entire project not for individual activities.

Term	Value
SV	$500,000
CV	-$550,000
SPI	1.15
CPI	0.87
EAC	$8.39 million
ETC	$8.39 million
VAC	-$1.09 million

Table A.2

From the above data from our project, we see the project is nicely ahead of schedule, as shown by our SPI greater than one and our positive SV. But the project is quite a bit over budget, as shown by our CPI of 0.87 and our negative CV of -$550,000. If spending continues at the present rate, we will run approximately a million dollars over budget. Something should be done to rein in costs.

The project's CPI and SPI provide much useful data with which to manage a project. Generally, a SPI or CPI between 0.9 and 1.2 can be ignored. This may be caused by normal fluctuations in a project. On the positive side, a CPI or SPI values between 1.2 and 1.3 should be evaluated when time permits. A CPI or SPI value greater than 1.3 should be looked into immediately. Sometimes when things look too good to be true, they just may be. On the negative side, CPI or SPI values between 0.8 and 0.9 should be evaluated as time permits. Between 0.8 and 0.6, they should be evaluated immediately as the project is certainly in trouble. CPI or SPI values less than 0.6 are a serious issue. These require immediate action to save the project. The cause of low values of CPI or SPI can be traced back to the activity or activities that are causing the issue by performing individual calculations on each activity.

Earned Value Management Formulas

AC = Actual Cost of the Work Performed

EV = Earned Value
- EV = Budgeted Cost of the Work Performed (BCWP)
- EV = % complete times BAC

PV = Planned Value
- PV = Budgeted Cost of the Work Scheduled (BCWS)
- PV = % of time expended times BAC

CV = Cost Variance
- CV = EV − AC

CPI = Cost Performance Index
- CPI = EV/AC

SV = Schedule Variance
- SV = EV − PV

SPI = Schedule Performance Index
- SPI = EV/PV

EAC = Estimate at Completion
- EAC = BAC/CPI − Same rate of spending
- EAC= AC + (BAC − EV)/CPI -- typical
- EAC= AC + ETC − fundamentally flawed
- EAC = AC + (BAC − EV) -- atypical

ETC = Estimate to Complete
- ETC = EAC − AC

VAC = Variance at completion
- VAC = BAC − EAC

BAC = Budget at Completion (Project budget)

Table A.3: Earned Value Formulae

Sample Risk Register

Project Title:	*Project Number:*

Risk Title:	
Risk Description:	

Risk Number:	*WBS Number:*
Risk Rating:	
Probability: Very Low/Low/Medium/High/Very High	
Impact: Very Low/Low/Medium/High/Very High	
Risk Ranking:	*Risk Owner:*
Risk Status: Occurred/Expired/Closed/Deleted	*Risk Status Date:*
Risk Trigger:	
Risk Response Actions:	*Was Response Effective?*

Contingency Plan:	Was Contingency Plan Effective?
Notes:	

Table B.1: Sample Risk Register

Cumulative (Single Tail) Probabilities of the Normal Distribution

Example: the area to the left of Z = 1.34 is found by following the left Z column down to 1.3 and moving right to the .04 column. At the intersection read .9099. The area to the right of Z = 1.34 is 1 - .9099 = .0901. The area between the mean (dashed line) and Z = 1.34 = .9099 - .5 = .4099.

z	.00	.01	.02	.03	.04
.0	.5000	.5040	.5080	.5120	.5160
.1	.5398	.5438	.5478	.5517	.5557
.2	.5793	.5832	.5871	.5910	.5948
.3	.6179	.6217	.6255	.6293	.6331
.4	.6554	.6591	.6628	.6664	.6700
.5	.6915	.6950	.6985	.7019	.7054
.6	.7257	.7291	.7324	.7357	.7389
.7	.7580	.7611	.7642	.7673	.7704
.8	.7881	.7910	.7939	.7967	.7995
.9	.8159	.8186	.8212	.8238	.8264
1.0	.8413	.8438	.8461	.8485	.8508
1.1	.8643	.8665	.8686	.8708	.8729
1.2	.8849	.8869	.8888	.8907	.8925
1.3	.9032	.9049	.9066	.9082	.9099
1.4	.9192	.9207	.9222	.9236	.9251
1.5	.9332	.9345	.9357	.9370	.9382
1.6	.9452	.9463	.9474	.9484	.9495
1.7	.9554	.9564	.9573	.9582	.9591
1.8	.9641	.9649	.9656	.9664	.9671
1.9	.9713	.9719	.9726	.9732	.9738
2.0	.9772	.9778	.9783	.9788	.9793
2.1	.9821	.9826	.9830	.9834	.9838
2.2	.9861	.9864	.9868	.9871	.9875
2.3	.9893	.9896	.9898	.9901	.9904
2.4	.9918	.9920	.9932	.9925	.9927
2.5	.9938	.9940	.9941	.9943	.9945
2.6	.9953	.9955	.9956	.9957	.9959
2.7	.9965	.9966	.9967	.9968	.9969
2.8	.9974	.9975	.9976	.9977	.9977
2.9	.9981	.9982	.9982	.9983	.9984
3.0	.9987	.9987	.9987	.9988	.9988
3.1	.9990	.9991	.9991	.9991	.9992
3.2	.9993	.9993	.9994	.9994	.9994
3.3	.9995	.9995	.9995	.9996	.9996
3.4	.9997	.9997	.9997	.9997	.9997

z	.06	.07	.08	.09
.0	.5239	.5279	.5319	.5359
.1	.5636	.5675	.5714	.5753
.2	.6026	.6064	.6103	.6141
.3	.6406	.6443	.6480	.6517
.4	.6772	.6808	.6844	.6879
.5	.7123	.7157	.7190	.7224
.6	.7454	.7486	.7517	.7549
.7	.7764	.7794	.7823	.7852
.8	.8051	.8078	.8106	.8133
.9	.8315	.8340	.8365	.8389
1.0	.8554	.8577	.8599	.8621
1.1	.8770	.8790	.8810	.8880
1.2	.8962	.8980	.8997	.9015
1.3	.9131	.9147	.9162	.9177
1.4	.9279	.9292	.9306	.9319
1.5	.9406	.9418	.9429	.9441
1.6	.9515	.9525	.9535	.9545
1.7	.9608	.9616	.9625	.9633
1.8	.9686	.9693	.9699	.9706
1.9	.9750	.9756	.9761	.9767
2.0	.9803	.9808	.9812	.9817
2.1	.9846	.9850	.9854	.9857
2.2	.9881	.9884	.9887	.9890
2.3	.9909	.9911	.9913	.9916
2.4	.9931	.9932	.9934	.9936
2.5	.9948	.9949	.9951	.9952
2.6	.9961	.9962	.9963	.9964
2.7	.9971	.9972	.9973	.9974
2.8	.9979	.9979	.9980	.9981
2.9	.9985	.9985	.9986	.9986
3.0	.9989	.9989	.9990	.9990
3.1	.9992	.9992	.9993	.9993
3.2	.9994	.9995	.9995	.9995
3.3	.9996	.9996	.9996	.9997
3.4	.9997	.9997	.9997	.9998

U. S. Navy Fleet Ballistic Missiles

Evolution of Fleet Ballistic Missiles

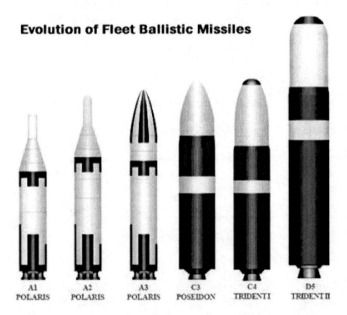

*Figure D.1: Evolution of U. S. Navy Fleet Ballistic Missile
(Courtesy of the U.S. Navy)*

U.S. Navy Fleet Ballistic Missile Statistics

	POLARIS (A1)	POLARIS (A2)	POLARIS (A3)
Length	28 feet	31 feet	32 feet
Diameter	54 inches	54 inches	54 inches
Weight	28,000 pounds	32,500 pounds	35,700 pounds
Powered Stages	2	2	2
Motor Case Materials	1st Stage - Low alloy steel 2nd Stage - Low alloy steel	1st Stage - Steel 2nd Stage - Glass Fiber	1st Stage - Glass Fiber [1] 2nd Stage - Glass Fiber [1]
Nozzles	4, each stage	4, each stage	4, each stage
Controls	Jetevators	1st Stage - Jetevators 2nd Stage - Rotating Nozzles	1st Stage - Rotating Nozzles 2nd Stage - Fluid Injection [2]
Propellant	Solid	Solid	Solid 1st Stage - Composite
Guidance	All Inertial	All Inertial	All Inertial
Range (nominal)	1,200 NM (1,380 SM)	1,500 NM (1,730 SM)	2,500 NM (2,880 SM)
Warheads	Nuclear	Nuclear	Nuclear

Figure D.2: Comparison of Fleet Ballistic Missiles (Courtesy of the U.S. Navy)

	POSEIDON (C3)	TRIDENT I (C4)	TRIDENT II (D5)
Length	34 feet	34 feet	44 feet
Diameter	74 inches	74 inches	83 inches
Weight	64,000 pounds	73,000 pounds	130,000 pounds (approx)
Powered Stages	2	3	3
Motor Case Materials	1st Stage - Glass Fiber 2nd Stage - Glass Fiber	All 3 Stages Kevlar/Epoxy	1st Stage - Graphite/Epoxy 2nd Stage - Graphite/Epoxy 3rd Stage - Kevlar/Epoxy
Nozzles	1, each stage	1, each stage	1, each stage
Controls	Single Movable Nozzle Actuated by a Gas Generator	Single Movable Nozzle Actuated by a Gas Generator	Single Movable Nozzle Actuated by a Gas Generator
Propellant	Solid 1st Stage - Composite	Solid Cross-Linked Double Base	Solid Nitrate Ester Plasticized Polyethylene Glycol
Guidance	All Inertial	Stellar and Inertial	Stellar and Inertial
Range (nominal)	2,500 NM (2,880 SM)	4,000 NM (4,600 SM)	> 4,000 NM (4,600 SM)
Warheads	Nuclear	Nuclear	Nuclear

Notes:

[1] First large ballistic missile to use glass motor case for all stages. (Small glass-fiber motor case had previously flown in Vanguard Program. POLARIS was the first large glass-fiber rocket motor case.)

[2] Devised and first flown by Navy in POLARIS development program.

Polaris Project Timeline

February 1955 - U. S. National Security Council report recommended ballistic missiles be developed for sea launch

August 1955 – Admiral Arleigh Burke appointed to Chief of Naval Operations, U. S. Navy.

November 8, 1955 – Admiral Raborn appointed Director Special Projects.

Summer of 1956 – the National Academy of Science Committee for Undersea Warfare was convened at Woods Hole (Project Nobska) at the request of CNO, Adm. Arleigh Burke. Dr. Bothwell, Head of Weapons Planning at China Lake Naval Station, addressed fleet ballistic missile (FBM) concepts. Dr. Bothwell, armed with the China Lake studies, proposed entirely different and smaller missile than the Jupiter Missile. His proposal met considerable resistance.

August 1956 – Dr. Edward Teller when questioned regarding the solid propellant Jupiter Missile being considered for the FBM, responded by asking "Why are you designing a 1965 weapon system with 1958 technology?" In his discussion, Teller verified what Bothwell had been telling them about advances in warhead technology.

September 4, 1956 – Dr. Teller submitted his warhead predictions to the AEC for certification. This validated the work at China Lake on the possibility of a smaller warhead size.

Labor Day weekend September 1956 – Bothwell and Witcher, a Weapons Planning Analysis, met with Captain Smith and Dr. Thompson at the SPO in Washington DC. The next day they informally briefed Adm. Raborn and staff. The concept of a small missile was immediately adopted.

October 1956 – Project Nobska results were presented to Adm. Burke and staff. It recommended a two stage solid fueled propellant missile system

November 1956 – The small missile concept was reviewed by the Scientific Advisory Committee

Late 1956 – Captain Smith appointed Technical Director, Special Project Office.

December 1956 – DOD approved a plan for shifting from the Army's Jupiter to the Navy's solid propellant missile system concept called 'Polaris'. Hence, the FBM Polaris was officially born.

January 1957 – SPO established a steering committee to study all aspects of the Polaris system design with a recommendation on the optimum size due 1 April 1957.

During this period of time, China Lake was completing its own independent study and issued a report (Project Mercury) on 28 February 1957. A smaller missile and warhead yield were proposed by China Lake, but the committee established a missile system with more conservative characteristics.

October 4, 1957 – Sputnik launched

Second week of October 1957 – U.S. Government restored funding to the Polaris Project.

November 3, 1957 – Sputnik 2 launched

December 6, 1957 – Vanguard Rocket explodes upon launch. This launch is televised on national television.

December 30, 1957 – The U.S. Navy decided to convert the U.S.S. Scorpion (SSN598) to the U.S.S. George Washington (SSBN 598).

January 17, 1958 – First test of launch of Polaris launch vehicle at Cape Canaveral, Florida.

January 31, 1958 – U. S. Navy Vanguard Rocket places Explorer 1 in orbit. This is the first successful United State orbital launch.

March 23, 1958 – First dummy test of Polaris submarine launch system at China Lake, CA.

July 22-23, 1958 – New W-47 Polaris warhead successfully tested at Bikini Atoll. Warhead exceeded expectations.

June 9, 1959 – U.S.S. George Washington Launched in Groton, Connecticut

September 22, 1959 – U.S.S. Patrick Henry launched in Groton, Connecticut.

November 20, 1959 – Polaris Missile launched from ship Observation Island off Cape Canaveral, Florida

December 30, 1959 – U.S.S. George Washington commissioned.

March 9, 1960 – Polaris Missile launched from Cape Canaveral, Florida.

July 20, 1960 – U.S.S. George Washington fires Polaris missile while submerged.

Timeline of the Polaris Project derived from Emme, E. Aeronautics and Astronautics: An American Chronology of Science and Technology in the Exploration of Space, 1915-1960 and Knemeyer, F. Concept Formulation of the Navy's FBM.

Glossary and Abbreviations

AC – Actual Costs to date used in Earned Value Management

AEC – United States Atomic Energy Commission

ARPANET - Advanced Research Projects Agency Network. A precursor to today's internet.

BAC – Budget at Completion

Cause and Effect Diagram – Diagrams that illustrate the causes of events. This type of diagram attempts to illustrate all known causes and sub-causes to an event or problem on a project. It is also known as a Fishbone Diagram or as an Ishikawa Diagram.

Checklist – an organized list of risks usually identified from earlier projects that is used to assist in risk identification in the current project.

Constraint – A condition that restricts a project's options or dictates project team actions.

Contingency Reserve – An amount of money set aside for accepted risks (known-unknown risks).

Cost Reimbursable – a contract in which the seller is paid for allowable expenses, such as labor and materials, and an additional payment for profit.

CPI – Cost Performance Index

CPM – Critical Path Method

Crash/crashing – a method to reduce a project's duration by adding more resources, such as money or personnel.

Critical Path – Those activities from start to finish that from the longest path through the project.

DARPA – Defense Advanced Research Projects Agency. This agency is part of the United States Department of Defense and is tasked with maintaining technological superiority of U.S. Military. Its purpose is also to prevent technological surprise from adversaries of the United States from harming the United States.

Delphi – A risk identification technique that uses a consensus of experts. This technique employs a facilitator to gather responses from the subject matter experts, correlate these responses, and redistribute them back to the experts. Subject matter experts participate anonymously in order to minimize bias and groupthink.

DOD – United States Department of Defense.

EAC – Estimate at Completion

EMV – Expected Monetary Value is the weighted expected value of a decision or outcome. The technique employs probability and impact associated with various possible outcomes to determine that decision or outcome's weighted value. This method assists in making better objective decisions.

EVM – Earned Value Management is a project management and control technique that objectively measures a project's performance.

EV – Earned Value

Expert Judgment – the use of individuals or groups that possess specialized knowledge, skills, or abilities in a specific task to be performed to provide input as to the methodology to be used to accomplish that task.

Failure Modes and Effect Analysis – A stepwise technique to identify all possible failures associated with a process, operation, system, product, or service. These failures are then grouped by classification and probability of occurrence.

FBM – Fleet Ballistic Missile Program

Fixed Price Contract – A type of contract that provides a firm fixed price for the work or services to be performed.

Fishbone Diagram – see Cause and Effect Diagram

FMEA – See Failure Modes and Effect Analysis

Force Field Analysis – A technique that explores forces for and forces against a given decision or project. This technique is very useful in identifying stakeholder related risks associated with a project.

Gold Platting – providing the customers more than the customers asked for or required as a way of enhancing customer satisfaction.

Groupthink – A term coined by Irving Janis that depicts a group making a faulty decision due to pressures within the group.

ICBM – Intercontinental Ballistic Missile

Ishikawa Diagram – See Cause and Effect Diagram

Managerial Reserve – An amount of money set aside for unknown-unknown risks.

Manhattan Project – Code name for the World War II project to develop the first atomic bomb.

NASA – National Aeronautics and Space Administration. NASA is an entity of the United States government.

Nominal Group – A decision making or information gathering technique, similar to Brainstorming, that requires all participants to actively participate.

Pareto Principle – Also known as the 80/20 rule. This principle states that 80% of all issues come from 20% of the causes. Hence, focusing on the 20% most important causes will eliminate 80% of one's issues on a project.

PERT – Program Evaluation and Review Technique

PMI – Project Management Institute, Newtown Square, PA

PMP – Project Management Professional Certification

PV – Planned Value

RACI – A chart that shows an individual's responsibility with respect to specific work on a project. The acronym stands for: Responsible, Accountable, Consult, and Inform.

RAM – Responsibility Assignment Matrix. A matrix based chart that marries roles and responsibilities with work elements on a project.

RBS – See Risk Breakdown Structure

Risk Breakdown Structure – a method to display risk categories and subcategories.

Risk Tolerance – that amount of risk the project stakeholders are comfortable with after considering the benefits and losses on a given project

Reserve Analysis – A method to establish a project's schedule, cost, and contingency reserves.

Residual Risk – A risk that remains after risk responses have been created and implemented.

Secondary Risk – A risk that arises as a direct result of a risk response.

Sensitivity Analysis – An analysis in which key assumptions or parameters are varied to determine their particular effect on overall project parameters and outcomes.

SINS – Ships Internal Navigation System. SINS is an internal navigation system utilizing accelerometers, computers, and gyroscopes to calculate a ship's position, speed, and orientation without the need for external references.

SLBM – Submarine Launched Ballistic Missile

SPI – Schedule Performance Index

SSN – Submersible ship nuclear powered. This is a United States Naval term for a nuclear powered attack submarine.

SSBN – Submersible Ship Ballistic Missile Nuclear Powered. This is a United States Naval Term for a nuclear powered submarine capable of launching ballistic missiles.

Standard Deviation – is the measure of how far a point varies from the statistical average or mean.

States of Nature – possible outcomes to a given situation of which only one of these outcomes can actually occur. Which outcome will actually occur is out the decision maker's control.

SV – Schedule Variance

SWOT – A planning method to evaluate a project's Strengths, Weaknesses, Opportunities, and Threats.

T & M – Time and Materials Contract

Tornado Diagram – A diagram used for sensitivity analysis. Variables that affect the project are displayed on the diagram. Variables with the greatest effect on project objectives are displayed on the top of the diagram as horizontal bars. As these variables move further down the diagram, their effect on the project decreases and their horizontal bar length decreases. Variables with the least effect are shown at the bottom of the diagram. This diagram provides a quick and easy method to identify those variables most important to project success.

Trigger – an indication or symptom that a risk has occurred or is about to occur.

Urgent Risks – Risks that may occur early in the project. These should be address before other risks, as they may occur before the normal risk management processes are completed.

USSR – Union of Soviet Socialist Republics also known as the Soviet Union. Most of this country today is known as Russia.

Variance – In statistics, the variance is the standard deviation squared.

VAC – Variance at Completion

Watch list – a list of risks upon which no action will be taken at this time to address. These risks are watched and monitored during project execution. Generally, these risks have a combination of low probability and/or low risk – making their significance and risk ranking low to the project.

WBS – Work Breakdown Structure

Workaround – a response to a negative risk that has occurred. This risk was not previously planned for in risk response planning. Workarounds are normally the result of an unknown-unknown risk occurring.

Z Score – a measure of distance of a sample from the mean measured in standard deviations.

References

Barkley, Bruce T. (2004). *Project Risk Management.* New York: McGraw-Hill.

Bird, Kai, Sherwin, Martin. (2005). *American Prometheus the Triumph and Tragedy of J. Robert Oppenheimer.* New York: Vintage Books.

Carbone, Thomas A. and Sampson Gholston. (2004). "Project manager skill development: A survey of programs and practitioners". *Engineering Management Journal,* 16(3): 10-16. Retrieved from http://www.tomcarbone.com/papers/Carbone-EMJSept04-PM.pdf

Emme, Eugene. (1961). *Aeronautics and Astronautics: An American Chronology of Science and Technology in the Exploration of Space, 1915-1960.* Washington, D.C.: NASA.

Garber, Steve. (2007). "Sputnik and the Dawn of the Space Age". National Atmospheric and Space Administration. Retrieved from http://history.nasa.gov/sputnik/.

German Submarine U-511. (n. d.) Retrieved form
    http://en.wikipedia.org/wiki/German_
    submarine_U-511.

Fact (n. d.). In Merriam-Webster Dictionary online. Retrieved
    form http://www.merriam-webster.com/netdict/
    fact.

Herken, Gregg (2002). *Brotherhood of the Bomb.* New York:
    Holt Paperbacks.

Hillson, David, & Simon, Peter. (2007). *Practical Project Risk
    Management: the ATOM Methodology.* Vienna,
    Virginia: Management Concepts.

Hillson, David. (2002). "The Risk Breakdown Structure
    (RBS) as an Aid to Effective Risk Management".
    Presented at the Fifth European Project
    Management Conference, PMI Europe 2002,
    Cannes, France.

Heldman, Kim. (2009). *PMP® Project Management Professional
    Exam Study Guide 5th Edition.* Hoboken, New
    Jersey: Wiley Publishing.

Heldman, Kim. (2005). *Project Manager's Spotlight on Risk
    Management.* Alameda, CA. SYBEX Inc.

Knemeyer, Franklin, (2003). *Concept Formulation of the Navy's
    FBM.* U.S. Naval Weapons Station, China Lake,
    California.

Knemeyer, Franklin, (1957). *Notes from Origin of Atlantis.*
    U.S. Naval Weapons Center, China Lake,
    California.

Meilinger, Phillip. (1989). "The Admirals' Revolt of 1949:
    Lessons for Today". *Parameters.* September,1989,
    pages 81-96.

National Science Foundation (ed.) (2010). "A Timeline of
    NSF History". National Science Foundation.
    Retrieved from http://www.nsf.gov/about/history/
    overview-50.jsp.

Polmar, Norman. (2003). "The Polaris a Revolutionary Missile
    System and Concept". United States Navy
    Historical Center. Retrieved from http://www.
    history.navy.mil/colloquia/cch9d.html.

Project Management Institute. (2008). *A Guide to the Project
    Management Body of Knowledge (PMBOK® Guide)
    4th Edition*. Newtown Square, PA: Project
    Management Institute.

Raborn, William. (n. d.). Retrieved from http://en.wikipedia.
    org/wiki/William_Raborn.

Royer, Paul. (2002). *Project Risk Management A Proactive
    Approach*. Vienna, Virginia: Management
    Concepts.

Sapolsky, Harvey. (n. d.) *The U.S. Navy's Fleet Ballistic Missile
    Program and Finite Deterrence*. Massachusetts
    Institute of Technology, Cambridge,
    Massachusetts.

Schuyler, John. (2001). *Risk and Decision Analysis in Projects
    2nd Edition*. Newtown Square, PA: Project
    Management Institute.

Sputnik 1 (n. d.). Retrieved from http://en.wikipedia.org/wiki/
    Sputnik_1.

Sputnik Crisis (n. d.). Retrieved from http://en.wikipedia.org/
    wiki/Sputnik_crisis.

Standish Group (2009, April 23). "New Standish Group
    report shows more project failing and less
    successful projects". Standish Group Retrieved

from http://standishgroup.com/newsroom/ chaos_2009.php.

Symptom (n. d.). In Merriam-Webster Dictionary online. Retrieved from http://www.merriam-webster.com/ netdict/symptom.

Tague, Nancy. (2004). The Quality Tool Box 2nd edition. Milwaukee, Wisconsin: ASQ Quality Press.

Van Atta, Richard. (n. d.) Fifty Years of Innovation and Discovery. Defense Advanced Research Projects Agency – United States Department of Defense. Retrieved from http://www.darpa.mil/Docs/ Intro_-_Van_Atta_200807180920581.pdf.

Watson, John (1998). "The Origin of the APL Strategic Systems Department". *John Hopkins Technical Digest*, v19 number 4.

Weir, Gary (n. d.). "Deep Ocean, Cold War". U. S. Naval Historical Center. Retrieved from http://www.navy. mil/navydata/cno/n87/usw/issue_7/deepocean. htm.

Wideman, R. M. (ed.) (1992). *Project & Program Risk Management: a Guide to Managing Project Risks & Opportunities.* Newtown Square, PA: Project Management Institute.

About the Author

D r. John Byrne, PMP is an Associate Professor at DeVry University and Senior Faculty at the Keller Graduate School of Management. Dr. Byrne has a Doctorate in Business Administration in Management from the University of Sarasota, a MBA degree from Wilmington University, and a BS in Nuclear Engineering.

John served ten years in the US Navy as a submarine nuclear reactor operator stationed on the ballistic missile submarine USS Nathanael Greene (SSBN 636) and the attack submarine USS Boston (SSN 703).

He is an outspoken advocate of project management, and has spoken at the Project Management Institute Leadership Sessions at PMI's Global Congress and has presented at PMI Global Congress Breakout Sessions in the past. Additionally, Dr. Byrne has also done presentations for the local chapters of PMI.

John is a consultant to industry on Project Management and has been a practitioner of project management for over 20 years. He has written books and articles for the Project Management Institute's Educational Foundation's K-12

Education Initiative and is on the Project Management Institute Educational Foundation's International Visioning Committee for K-12 Project Management Education.

John lives with his family in Northern Delaware.

Did you like this book?

If you enjoyed this book, you will find more interesting books at

www.MMPubs.com

Please take the time to let us know how you liked this book. Even short reviews of 2-3 sentences can be helpful and may be used in our marketing materials. If you take the time to post a review for this book on Amazon.com, let us know when the review is posted and you will receive a free audiobook or ebook from our catalog. Simply email the link to the review once it is live on Amazon.com, with your name, and your mailing address—send the email to orders@mmpubs. com with the subject line "Book Review Posted on Amazon."

If you have questions about this book, our customer loyalty program, or our review rewards program, please contact us at info@mmpubs.com.

Project Lessons from The Great Escape (Stalag Luft III)

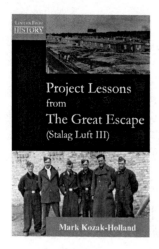

While you might think your project plan is perfect, would you bet your life on it?

In World War II, a group of 220 captured airmen did just that – they staked the lives of everyone in the camp on the success of a project to secretly build a series of tunnels out of a prison camp their captors thought was escape proof.

The prisoners formally structured their work as a project, using the project organization techniques of the day. This book analyzes their efforts using modern project management methods and the nine knowledge areas of the *Guide to the Project Management Body of Knowledge* (PMBoK).

Learn from the successes and mistakes of a project where people really put their lives on the line.

ISBN: 9781895186802 (paperback)

Also available in ebook formats. Order from your local bookseller, Amazon.com, or directly from the publisher at **http://www.mmpubs.com/escape**

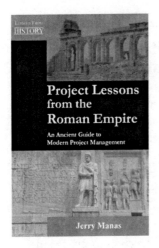

Project Lessons from the Roman Empire: An Ancient Look at Modern Project Management

The leaders of the Roman Empire established many of the organizational governance practices that we follow today, in addition to remarkable feats of engineering using primitive tools that produced roads and bridges which are still being used today, complex irrigation systems, and even "flush toilets." Yet, the leaders were challenged with political intrigue, rebelling team members, and pressure from the competition. How could they achieve such long-lasting greatness in the face of these challenges?

In this new addition to the Lessons from History series, join author Jerry Manas as he takes you on a journey through history to learn about project management the Roman way. Discover the 23 key lessons that can be learned from the successes and failures of the Roman leadership, with specific advice on how they can be applied to today's projects.

Looking at today's hottest topics, from the importance of strategic alignment for your projects through to managing transformational change and fostering work/life balance while still maintaining overall performance, you'll find that the Romans already faced-and conquered-these challenges two thousand years ago. Read this intriguing book to learn how they did it.

ISBN: 9781554890545 (paperback)

Also available in ebook formats. Order from your local bookseller, Amazon.com, or directly from the publisher at **http://www.mmpubs.com/**

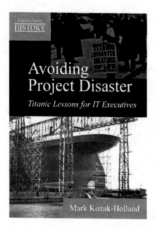

Avoiding Project Disaster: Titanic Lessons for IT Executives

Imagine you are in one of *Titanic's* lifeboats. As you look back at the wreckage, you wonder what could have happened. What were the causes? How could things have gone so badly wrong?

Titanic's maiden voyage was a disaster waiting to happen as a result of the compromises made in the project that constructed the ship. This book explores how modern executives can take lessons from a nuts-and-bolts construction project like *Titanic* and use those lessons to ensure the right approach to developing online business solutions.

Avoiding Project Disaster is about delivering IT projects in a world where being on time and on budget is not enough. You also need to be up and running around the clock for your customers and partners. This book will help you successfully maneuver through the ice floes of IT management in an industry with a notoriously high project failure rate.

ISBN: 9781895186734 (paperback)

Also available in ebook formats. Order from your local bookseller, Amazon.com, or directly from the publisher at **http://www.mmpubs.com/disaster**

Titanic Lessons for IT Projects

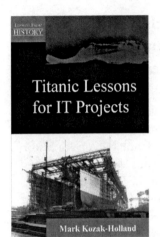

Titanic Lessons for IT Projects analyzes the project that designed, built, and launched the ship, showing how compromises made during early project stages led to serious flaws in this supposedly "perfect ship." In addition, the book explains how major mistakes during the early days of the ship's operations led to the disaster. All of these disasterous compromises and mistakes were fully avoidable.

Entertaining and full of intriguing historical details, this companion book to *Avoiding Project Disaster: Titanic Lessons for IT Executives* helps project managers and IT executives see the impact of decisions similar to the ones that they make every day. An easy read full of illustrations and photos to help explain the story and to help drive home some simple lessons.

ISBN: 9781895186260 (paperback)

Also available in ebook formats. Order from your local bookseller, Amazon.com, or directly from the publisher at **http://www.mmpubs.com/titanic**

Churchill's Adaptive Enterprise: Lessons for Business Today

This book analyzes a period of time from World War II when Winston Churchill, one of history's most famous leaders, faced near defeat for the British in the face of sustained German attacks. The book describes the strategies he used to overcome incredible odds and turn the tide on the impending invasion. The historical analysis is done through a modern business and information technology lens, describing Churchill's actions and strategy using modern business tools and techniques.

Aimed at business executives, IT managers, and project managers, the book extracts learnings from Churchill's experiences that can be applied to business problems today. Particular themes in the book are knowledge management, information portals, adaptive enterprises, and organizational agility.

ISBN: 9781895186192 (paperback)

Also available in ebook formats. Order from your local bookseller, Amazon.com, or directly from the publisher at **http://www.mmpubs.com/churchill**